ANALOGIES IN OPTICS AND MICRO ELECTRONICS

T0321142

Analogies in Optics and Micro Electronics

Selected Contributions on Recent Developments

Edited by

Willem van Haeringen

Eindhoven University of Technology, Eindhoven, The Netherlands

and

Daan Lenstra

Eindhoven University of Technology and University of Leiden, The Netherlands

KLUWER ACADEMIC PUBLISHERS
DORDRECHT / BOSTON / LONDON

Library of Congress Cataloging in Publication Data

Analogies in optics and micro electronics : selected contributions on
  recent developments / edited by Willem van Haeringen and Daan
  Lenstra.
      p.   cm.
  ISBN 0-7923-0708-9 (alk. paper)
  1. Optics. 2. Microelectronics.   I. Haeringen, Willem van, 1933-
. II. Lenstra, Daan, 1947-   .
TA1675.A49   1990
535--dc20   90-4175

ISBN 0-7923-0708-9

Published by Kluwer Academic Publishers,
P.O. Box 17, 3300 AA Dordrecht, The Netherlands.

Kluwer Academic Publishers incorporates
the publishing programmes of
D. Reidel, Martinus Nijhoff, Dr W. Junk and MTP Press.

Sold and distributed in the U.S.A. and Canada
by Kluwer Academic Publishers,
101 Philip Drive, Norwell, MA 02061, U.S.A.

In all other countries, sold and distributed
by Kluwer Academic Publishers Group,
P.O. Box 322, 3300 AH Dordrecht, The Netherlands.

Printed on acid-free paper

Printed in the Netherlands

CONTENTS

PREFACE

This book gives an account of a number of recent developments in two different subfields of research, optics and micro–electronics. The leading principle in presenting them together in one book is the striking similarity between a variety of notions in these two research areas. We mention in this respect tunneling, quantum interference and localization, which are important concepts in quantummechanics and more specifically in condensed matter physics. Miniaturization in solid state engineering has led to new phenomena in which these concepts play their significant roles. As it is the wave character of electrons which is strongly emphasized in these phenomena one's attention is quite naturally directed to the field of optics in which the above quantum–mechanical notions all seem to have their direct classical wavemechanical counterparts.

Both micro–electronics and optics have been and still are in a mode of intensifying activity. The possibilities to technically "translate" devices developed within one research field to similar devices in the other field are strongly increasing. This opens, among other things, a door leading to "quantummechanics" on a macroscopic scale with visible light under relatively easily accessible experimental conditions, or to "wave optics" in the domain of solid state physics. Thinking in terms of analogies is important anyhow, but it is especially the cross–fertilization between optics and micro–electronics which according to the editors will lead to deepened insights and a new type of technology.

We have chosen for a presentation in which a restricted number of selected researchers in both fields sketch developments to which they have significantly contributed themselves. This will make the book of utmost importance for workers in the respective fields. Furthermore, the didactical skills of the contributing authors guarantee accessibility of the book to a much broader group of readers, e.g. graduate students, physicists and electrical engineers who are not particularly specialized in this area, but wish to obtain an overview. The book contains fifteen presentations which we have classified into three groups: I. principles and basic concepts, II. coherent optics and III. coherent electronics. Apart from giving an important collection of new articles on recent developments in optics and micro–electronics, the objective of the editors has been to demonstrate the apparent potentialities of thinking in terms of analogies.

Many interesting other analogies such as, e.g., between superconducting and optical devices are not dealt with in the present volume. Also several important earlier developments in the field of electron beam "optics" and X–ray diffraction in crystals have not been covered. For a book of this set up and size it is impossible to give a complete account. So, the outline of this first edited work on analogies in optics and micro–electronics reflects very much the opinions and preferences of the editors.

We would like to express our sincere gratitude to Ria Coopmans–van Basten for the secretarial and technical support she gave with great commitment to the book. Special thanks also to Ria Groenendijk and Brigitte Senden for their assistance at critical moments during the final stage of the project.

May this book serve to intensify the "analogy" kind of approach, thus leading to even more cross–fertilization and if possible new fascinating results in the borderline between two important technologies in the fields of optics and micro-electronics.

Eindhoven, Spring 1990

Willem van Haeringen
Daan Lenstra

PART I

PRINCIPLES AND BASIC CONCEPTS

PLAYING WITH ELECTRONS AND PHOTONS IN RINGS

Daan Lenstra and Willem van Haeringen

A "one–dimensional" ring configuration is a simple non–trivial system, particularly suited for a theoretical study of wave propagation dynamics under the influence of an external driving force. This ring system, although initially designed as an academic play field with non–interacting model electrons, becomes most relevant and experimentally accessible in the case of real–world optics.

1. INTRODUCTION

A ring is a very interesting system which combines two useful properties in an intriguing if not controversial way: it is a finite system with regard to its dimensions but endless for waves or particles traveling around and around. In optics both the ring resonator and the ring laser have been thoroughly investigated which has led to many applications related to their subtle and rich phenomenology. We mention the laser gyroscope [1], bistability effects in ring lasers [2] and, more recently, π–phase jumps [3] and optical bandstructures in ring resonators [4], the latter item being adressed to in the chapter by Woerdman and Spreeuw (Chap.9).

In submicron electronics the ring system has been used for measuring Aharonov–Bohm type magneto–resistance oscillations [5]. The chapter by Van Haesendonck is devoted to this and other nonlocal effects (Chap.11). Superconducting rings with Josephson junctions have been thoroughly investigated [6,7]. For the theoretical physicist the ring has become an important model system for studying fundamental properties of electron dynamics and electrical conduction as it avoids the necessity of introducing ad hoc boundary conditions which are to reflect externally controlled fluxes of incoming or outgoing particles (or waves).

On the one hand, a ring measures precisely one primitive cell, but on the other hand it is an infinite periodic system as well. Particles or waves which are traveling around in the ring can be accelerated due to an external force. Furthermore, they can be scattered (in fact, only *back* scattered in a 1D ring) and thus a nontrivial dynamics can be created. We will consider only elastic scattering, i.e., described by a certain structure in the potential energy function (electrons) or refractive index (photons). For these scattering processes the coherence of the waves before, during and after scattering is fully conserved. This is an interesting limiting situation in case of electrons because it resembles the situation at very low temperatures, where the increasing amount of experimental evidence shows that electrical conduction cannot be fully described anymore in terms of diffusion of carriers, but should rather be treated in terms of coherent wave propagation [8]. The chapters by Van Haesendonck (Chap.11), Büttiker (Chap.12) and van Houten and Beenakker (Chap.13) deal with this regime, while also the concept of resonant tunneling, studied by Eaves in Chap.14, implies coherent propagation of charge

3

W. van Haeringen and D. Lenstra (eds.), Analogies in Optics and Micro Electronics, 3–19.
© 1990 *Kluwer Academic Publishers. Printed in the Netherlands.*

carriers.

One could argue that analogies between electrons and light can best be studied on the level of the Dirac and Maxwell equations. However, it is well known that the physics of electrons in modern semiconducting materials and structures can succesfully be described within a Schrödinger–equation framework. It would therefore be extremely instructive and useful if existing analogies between certain aspects of optics and micro–electronics can be discussed within the framework of a Schrödinger equation covering both cases. Therefore, in Sec.2 we take up that challenge and arrive at the conclusion that photons can indeed adequately be dealt with by means of a non–relativistic Schrödinger equation.

In Sec.3 we use the ring configuration in order to present a unified treatment of wave dynamics in response to a slowly varying external perturbation. We find general properties of the eigenstates and corresponding eigenvalues and we treat the dynamical response in the adiabatic approximation. In Sec.4 we discuss the experimental feasibility of Bloch oscillations in a ring. By estimating the various time scales involved in such an experiment, we conclude that the best opportunities for observing Bloch oscillations are found in the optical ring system.

Non–adiabatic response, involving transitions between energy bands, is discussed in Sec.5. In this regime we can study, at least theoretically, the fundamental problem of coherent (i.e. temperature $T = 0K$) wave dynamics due to relatively large driving fields, without performing any of the usual approximations such as linear–response theory or Boltzmann–equation approach. In the case of electrons, the solution of the time–dependent Schrödinger equation turns out to be chaotic, which is related to the fact that the time–dependent phases (modulo 2π) in the one–electron wave function generally don't repeat themselves after each roundtrip in a regular or commensurate way. This gives the system self-randomizing properties due to pseudo–stochastics, that is, on short time scales. The intrinsic coherence of the multiply scattered waves becomes noticeable on longer time scales as a weak form of localization.

2. A SCHRÖDINGER EQUATION FOR PHOTONS

Since one of our aims is to exploit the analogy between electron waves and optical waves, it would be beneficial to deal with one and the same equation valid for either case. Confining ourselves to the rich variety of electron propagation phenomena which are succesfully described by a Schrödinger equation, for which the solutions and their properties are well known, we want to investigate the possibility of describing photons by means of a Schrödinger equation. In order to be able to study dynamical properties, it has to be a time–dependent Schrödinger equation. The derivation will quite naturally lead to the quantity which plays the role of *potential*; this will make it possible and easy to "translate" many typical quantummechanical results directly into the field of optics [9].

Our starting point is the wave equation for the electric–field component E in a configuration involving dielectric materials,

$$\frac{1}{c^2}\frac{d^2E}{dt^2} - \nabla^2 E = -\frac{1}{\epsilon_0 c^2}\frac{d^2P}{dt^2} . \tag{1}$$

For simplicity we consider the scalar case only; it is possible, however, to include the vector nature of the light and draw a parallel between the polarization of light and spin polarization of electrons. In fact, this is the topic discussed by Bhandari in Chap.5. The quantity P in the right hand side of (1) is the induced dipole moment density which quite often, but certainly not always, has a linear functional dependence on E. In vacuum, $P \equiv 0$, but in a material medium $P \neq 0$.

The characteristics of any Schrödinger equation are:

(i)   it is a first–order differential equation with respect to time;

(ii)   it has a kinetic energy term (second order derivative with respect to the space variable) which contains an inertial mass;

(iii)   it has a potential energy term.

By employing a weak signal approximation for the electric field and the dipole moment density we can indeed arrive at all three above items. In fact, the procedure is somewhat similar to deriving the Schrödinger equation for electrons from the Dirac equation. In the latter procedure one considers small energy deviations from the relativistic rest mass energy; in the case of light we consider small frequency deviations from a conveniently chosen fixed optical frequency. In this spirit we write for the fields E and P

$$E(r,t) = \mathscr{E}(r,t)e^{-i\omega_0 t} + c.c. \, ,$$

$$P(r,t) = \mathscr{P}(r,t)e^{-i\omega_0 t} + c.c. \, , \tag{2}$$

with ω_0 a conveniently chosen optical frequency such that the time dependence in $\mathscr{E}$ and $\mathscr{P}$ is slow enough to neglect the second–order time derivatives of $\mathscr{E}$ and $\mathscr{P}$ when (2) is substituted in (1). The resulting equation can be written in the form

$$i\hbar \frac{\partial \mathscr{D}}{\partial t} = \frac{-\hbar^2 \nabla^2}{2m_0}(\epsilon_0 \mathscr{E}) - \tfrac{1}{2}\hbar\omega_0 \mathscr{D}, \tag{3}$$

where we have introduced the dielectric displacement

$$\mathscr{D}(r,t) = \epsilon_0 \mathscr{E}(r,t) + \mathscr{P}(r,t) \tag{4}$$

and where $m_0 \equiv \hbar\omega_0/c^2$ plays the role of photon mass. In the simple case of propagation in a homogeneous dielectric with *constant* relative permittivity ϵ_r^0, we have $\mathscr{D} = \epsilon_0\epsilon_r^0 \mathscr{E}$, and (3) can be written as

$$i\hbar \frac{\partial \mathscr{D}}{\partial t} = \frac{-\hbar^2 \nabla^2}{2\epsilon_r^0 m_0} \mathscr{D} - \tfrac{1}{2}\hbar\omega_0 \mathscr{D}, \tag{5}$$

i.e. a Schrödinger equation with "effective" mass $\epsilon_r^0 m_0$ and constant potential $-\tfrac{1}{2}\hbar\omega_0$. Note that our mass m_0 agrees with the general Einstein relation $E = m_0 c^2$ as well as with the identification of $\hbar\omega_0$ with the energy of the photon in vacuum ($\epsilon_r^0 = 1$). On the other hand, it is only *half* rather than the *full* photon energy which shows up in the potential in the r.h.s. of (5). This is related to the fact that the total photon energy $\hbar\omega_0$ is equally partitioned between the *electric* and *magnetic* components of the electromagnetic field.

In a more general case we may write

$$\mathscr{P}(r,t) = \epsilon_0(\epsilon_r^0 - 1)\mathscr{E}(r,t) + \mathscr{P}'(r,t), \tag{6A}$$

where $\mathscr{P}'$ is any polarization contribution not yet accounted for by ϵ_r^0. Note that it is always possible to relocate linear polarization response taken from $\mathscr{P}'$ in ϵ_r^0 and vice versa; as we will see, this corresponds to renormalizing the photon mass. In actual cases it is advisable to choose ϵ_r^0 in such a way that the remaining part $\mathscr{P}'$ is minimized in some sense. From (6A) and (4) we find

$$\epsilon_0 \mathscr{E} = (\mathscr{D} - \mathscr{P}')/\epsilon_r^0 , \tag{6B}$$

which yields after substitution in (3) the following Schrödinger equation for $\mathscr{D}$:

$$i\hbar \frac{\partial \mathscr{D}}{\partial t} = \frac{-\hbar^2 \nabla^2}{2\epsilon_r^0 m_0} \mathscr{D} + V \mathscr{D}, \tag{7A}$$

where the potential operator is given by

$$V\mathscr{D} = -\tfrac{1}{2}\hbar\omega_0 \left(\mathscr{D} - \frac{c^2 \nabla^2}{\epsilon_r^0 \omega_0^2} \mathscr{P}' \right) . \tag{7B}$$

In general, the "potential" V must be considered a time–dependent, non–local and non–linear pseudo–potential, but such potentials, or as they are also called, self–energy functions are well known in the physics of many–particle systems. At this point we should also mention a fundamental difference between electrons and photons. Electrons are fermions and hence, subject to the Pauli exclusion principle, whereas photons are bosons for which there is no such exclusion principle. This difference is manifest in many–particle systems. In fact, in some respects photons look more like Cooper pairs with the coherent optical field as the analogue of the macroscopic superconducting wave function, but on the level of single–particle descriptions photons can be strikingly similar to electrons, as we will discuss in the following sections.

It is only under special circumstances that the potential operator V in (7B) reduces to an ordinary potential, i.e. an r–dependent function. As an example of the latter, consider the case in which $\mathscr{P}' = \epsilon_0[\epsilon_r(r) - \epsilon_r^0] \mathscr{E}$ with $\epsilon_r(r)$ a slowly varying function over distances of the order of the wavelength. $\mathscr{P}'$ can be expressed in terms of $\mathscr{D}$ as $\mathscr{P}' = [\epsilon_r(r) - \epsilon_r^0] \mathscr{D}/\epsilon_r(r)$. Using $\nabla^2 \mathscr{P}' \simeq -\omega_0^2 \epsilon_r(r) \mathscr{P}'/c^2$, we can approximate the potential in (7B) as

$$V(r) = -\tfrac{1}{2}\hbar\omega_0 \epsilon_r(r)/\epsilon_r^0 . \tag{8}$$

It must always be remembered, when applying (5) or (7A) in stationary situations, that the corresponding *time–independent* Schrödinger equation, i.e. $-\hbar^2/(2\epsilon_r^0 m_0)\nabla^2 \mathscr{D} + V\mathscr{D} = E\mathscr{D}$, is only valid so long as the energy E is close to zero, that is $|E| \ll \hbar\omega_0$, since otherwise the slowly varying envelope approximation (SVEA) for $\mathscr{E}$ and $\mathscr{P}$ would not be valid. Related to this, the SVEA is responsible for introducing a *parabolic* dispersion relation for photons, i.e. $E(k) = \hbar^2 k^2/(2\epsilon_r^0 m_0) - \tfrac{1}{2}\hbar\omega_0$, whereas we know that it should be linear with $dE/dk = \hbar c/(\epsilon_r^0)^{1/2}$. However, the difference between the parabolic and linear dispersion is hardly noticeable due to the relatively high kinetic energy ($\simeq \tfrac{1}{2}\hbar\omega_0$) associated with the above requirement $|E| \ll \hbar\omega_0$. Realizing that $|k|$ should not be very different from $k_0 = (\epsilon_r^0)^{1/2}\omega_0/c$, we observe that the parabolic dispersion relation in first approximation is linear with the correct shape $dE/dk = \hbar c/(\epsilon_r^0)^{1/2}$.

The above restriction on photon energies is absent in case of electrons. One must therefore be careful in translating properties based on the parabolic band shape directly into the optical domain. We note, however, that special configurations exist in which the photon dispersion relation *is* parabolic, for instance, when operating at frequencies close to an optical bandgap created by a Bragg reflection in an optical crystal of the kind described by Yablonovitch in Chap.8.

3. ADIABATIC RESPONSE IN A RING

In this section we will study the response of waves in a circular ring to "slow" dynamical perturbations due to an externally applied field. For the *electron* case we specifically think of a driving electric field $F(t)$ along the ring caused by a time–dependent magnetic flux $\Phi(t)$ through the ring area. Then, if the variable x measures the position along the ring, the Schrödinger equation for this situation is

$$i\hbar \frac{\partial \Psi(x,t)}{\partial t} = \left\{ \frac{1}{2m} \left[\frac{\hbar}{i} \frac{\partial}{\partial x} + \frac{e\Phi(t)}{L} \right]^2 + V(x) \right\} \Psi(x,t) , \qquad (9)$$

with L the ring circumference and $V(x)$ the potential energy of the electron. Single–valuedness of the wave function implies the periodic boundary condition

$$\Psi(x+L,t) = \Psi(x,t). \qquad (10)$$

Eq.(9) shows up in many more physical situations than the above–described one. A similar equation applies when there is no flux, but instead, the ring rotates with angular frequency Ω around the axis through the center and perpendicular to the rings face, so that each point of the ring will have velocity $v = L\Omega/2\pi$. This situation is adequately described by (9) if we replace $e\Phi(t)/L$ by $mL\Omega(t)/2\pi$. In fact, we would then be dealing with the Sagnac effect for non–relativistic particles [10]. We can easily project this onto the *optical* domain by using the Schrödinger equation for light, extended by the rotation–induced momentum contribution, yielding

$$i\hbar \frac{\partial \mathscr{D}}{\partial t} = \left\{ \frac{1}{2m_0} \left[\frac{\hbar}{i} \frac{\partial}{\partial x} + m_0 L\Omega(t)/2\pi \right]^2 + V(x) \right\} \mathscr{D}. \qquad (11)$$

Here we recall that the photon mass m_0 is given by $m_0 = \hbar\omega_0/c^2$ and $V(x)$ is the effective optical potential. In view of the perfect analogy between (9) and (11), many results obtained from (9) for electrons can immediately be taken over to optical waves in the rotating ring configuration.

If $\Phi(t)$ in (9) changes sufficiently slowly in time the adiabatic approximation applies and its results can serve as the starting point for obtaining the exact solution. In the adiabatic approximation we treat Φ as a constant, independent of time and we determine the Φ–dependent energy spectrum by solving the stationary Schrödinger equation

$$\left\{ \frac{1}{2m} \left[\frac{\hbar}{i} \frac{\partial}{\partial x} + \frac{e\Phi}{L} \right]^2 + V(x) \right\} u_n(\Phi,x) = E_n(\Phi)u_n(\Phi,x) , \qquad (12)$$

subject to the periodic boundary condition (10).

One easily verifies that the functions $\psi_n(\Phi,x)$ which are related to the $u_n(\Phi,x)$

through

$$\psi_n(\Phi,x) = e^{i\frac{e}{\hbar}\Phi\frac{x}{L}} u_n(\Phi,x) , \tag{13}$$

satisfy the unperturbed Schrödinger equation

$$\left[\frac{-\hbar^2}{2m}\frac{\partial^2}{\partial x^2} + V(x)\right]\psi_n(\Phi,x) = E_n(\Phi)\psi_n(\Phi,x) , \tag{14}$$

subject to the Bloch–type boundary condition

$$\psi_n(\Phi,x+L) = e^{i\frac{e}{\hbar}\Phi} \psi_n(\Phi,x) . \tag{15}$$

The phase reproduction factor in (15) clearly involves the Aharonov–Bohm phase $2\pi\Phi/\Phi_0$ where Φ_0 is the fundamental flux quantum,

$$\Phi_0 = h/e . \tag{16}$$

The Schrödinger problem (14) is nothing but the problem of determining the bandstructure of a one–dimensional periodic lattice with periodicity L and potential $V(x)$.

In the optical analogue described by (11) the role of the Aharonov–Bohm phase is taken over by $\Phi_s = m_0L^2\Omega/(2\pi\hbar)$ which can also be expressed as $\Phi_s = 2\omega_0\Omega A/c^2$, with $A \equiv L^2/(4\pi)$ the area of the ring. The quantity Φ_s is the extra roundtrip phase due to the rotation collected by a light wave of frequency ω_0. Since the counter propagating wave will have an equal but opposite contribution, we obtain for the difference in roundtrip time for both waves

$$\Delta T = 2\Phi_s/\omega_0 = 4\Omega A/c^2 , \tag{17}$$

which is the well–known Sagnac effect [11].

The above–derived Sagnac phase $\Phi_s = m_0L^2\Omega/h$ invites to the introduction of an optical analogue for the fundamental flux quantum, by writing

$$\Phi_s = \frac{m_0}{\hbar} C, \tag{18}$$

with $C = 2A\Omega$, known as the circulation. The fundamental circulation quantum for photons with mass m_0 is thus, in full analogy with (16), given by

$$C_0 = \frac{h}{m_0} . \tag{19}$$

We note that the optical quantum C_0 is less fundamental than the electronic flux quantum (16) in the sense that photons of all possible frequencies (mass) exist, whereas there is only one single electron charge known to us.

The Aharonov–Bohm phase present in (15) is one example of a phase

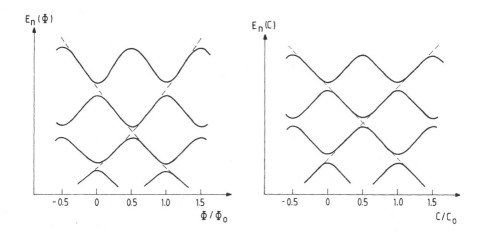

Fig.1   Energy–dispersion bands in a ring with elastic scattering. (a) Electron energies vs. normalized flux Φ/Φ_0 where $\Phi_0 = h/e$. (b) Photon energies vs. normalized circulation C/C_0 where $C_0 = h/m_0$ (see (19)).

anholonomy, that is, non–reproductivity of the phase, which is a fundamental concept in the theory of the Berry phase [12]. For an extensive discussion on this topic as well as the various Berry phases in optics we refer to the chapter by Chiao (Chap.10). We note that, in view of the above–derived analogy between the Aharonov–Bohm phase and the Sagnac phase the latter is also a Berry phase [13].

The question of the adiabatic (i.e., sufficiently slow) dynamics is in principle solved now for either system: The bandstructure functions $E_n(\Phi)$ or $E_n(\Omega)$ provide the clue to this, in crystal physics referred to as, *semiclassical* dynamics. The full analogy with the semiclassical model for Bloch particles in periodic configurations tells us immediately that the group velocity v_g is given by

$$v_g = \frac{\partial E_n}{\partial p}, \tag{20}$$

where p is a momentum, $p = e\Phi/L$ for electrons and $p = m_0 L\Omega/(2\pi)$ for photons. A typical bandstructure is sketched in Fig.1a for low–energy electrons and in Fig.1b for photons. In Fig.1a we recognize the underlying parabolic dispersion relation, where the gaps are due to coherent backscattering associated with a certain structure in the potential. In the limit of very strong backscattering the bands will tend to become flat. The optical case is similar to the electronic case at sufficiently high energies; the underlying free–photon bands are linear. Coherent backscattering in an optical ring also leads to gaps.

For weak backscattering the energy gap is in good approximation [14] given by

$$E_g = 2\hbar \left(vW/L\right)^{\frac{1}{2}}. \tag{21}$$

where v is the particle velocity in the absence of backscattering and W is the *elastic* backscattering rate. Since W equals the number of roundtrips per unit time (v/L) multiplied by the reflectivity per roundtrip (R), we can express E_g in the form $E_g = 2\hbar(v/L)R^{1/2}$, which for an optical band structure can be rewritten in terms of the frequency gap $\omega_g = E_g/\hbar$ in the appealing form

$$\omega_g = \frac{2c}{L}(R)^{1/2}.\tag{22}$$

This relationship has recently been confirmed by experiment [4].

4. DYNAMIC RESPONSE TO SLOW PERTURBATIONS; BLOCH OSCILLATIONS

There appears to be no consensus about what precisely is meant by Bloch oscillations. This is especially true when not dealing with electrons in an infinitely long, or at least very long crystal, but also the latter case leaves room for different viewpoints. However, all interpretations have in common that the application of a constant force field leads to a periodically oscillating response.

In case of electrons in a crystal the bandstructure is given in terms of periodic functions $E_n(k)$ in k-space; k is called the Bloch wave vector. These functions have the periodicity of the reciprocal lattice and application of a constant electric field F will cause the Bloch wave vectors to change according to $dk/dt = -eF/\hbar$, thus maintaining a uniform flow in k-space. In the semiclassical (adiabatic) approximation this flow causes, at least in one dimension, via (20) with $p = \hbar k$ a time–periodic current in real space with period $t_p = \hbar \Delta K/eF$ where ΔK is the Brillouin–zone width. In more than one dimension this would still be true if the field is directed along one of the principal symmetry axes of the crystal. The time t_p has the order of magnitude h/eV_p, where V_p is the potential drop over one primitive cell. With $V_p = 10^{-7}$ Volt we have $t_p \sim 4 \times 10^{-6}$ s which is typically a very long time for maintaining coherence of the wave function as is required in a Bloch–oscillation experiment. This relatively long coherence time is of course the reason for not having observed Bloch–oscillations in crystals, not even at very low temperatures.

One thinkable way to overcome the coherence problem is to reduce t_p, e.g. by increasing the linear dimension of the cell. Using, for instance, a periodicity length of 10 μm and a voltage difference of 10^{-2} Volt, we find $t_p \sim 4 \times 10^{-13}$ s. Coherence times exceeding this value by several orders of magnitude can easily be reached in a two–dimensional electron gas (2DEG) formed at the interface of two semiconductors [15]. However, the reproduction of this 10 μm cell over macroscopic distances will be technologically hard if not impossible. An obvious alternative one could think of is to implement in the 2DEG a ring with 10 μm circumference. In that case, no matter what kind of random impurities (elastic scattering centers) the ring may contain, the periodicity of the system will be perfect! Of course, the experiment should be carried out at sufficiently low temperature as we don't want too much inelastic scattering reducing our coherence time. For $T < 1$K it is not unreasonable to expect coherence times of the order 10^{-10} s, but we will show that this is still not sufficient for the Bloch–oscillation experiments.

Let us assume a bandgap E_g equal to 5 μeV and a Fermi velocity $v_F = 10^5$ m/s. In order to observe one period of a Bloch oscillation we must sweep the magnetic field B within the coherence time through at least one fundamental flux quantum,

$\Delta B = 4\pi\Phi_0/L^2 \simeq 0.5$ mT. With a coherence time $t_c = 100$ ps we find $\Delta B/\Delta t \geq$ 5 MT/s. On the other hand, in order to avoid Zener tunneling through the gap we must have[1] $F \equiv (1/L)d\Phi/dt << E_g^2/(2\pi e\hbar v_f)$ with v_f the Fermi velocity. In terms of the magnetic field rate of change this implies $\Delta B/\Delta t << 60$ KT/s. Therefore, we clearly have two conflicting conditions.

Let us now turn to an optical ring such as can be realized by using e.g. an optical fiber. As we will show, the prospects for observing Bloch oscillations in the optical domain are better than for the electric system. In fact, the experimental evidence for the occurence of optical Bloch oscillations has recently been reported by Spreeuw et al. [16]. In their experiment the driving force was not delivered by the rate of change in rotation frequency $d\Omega/dt$ (the ring was not rotating at all), but by simulating the effect of rotation via the Faraday effect influencing the polarization of the light [4]. It has been shown that the adiabatic properties of the system thus obtained are fully equivalent to those of the rotating ring [17]. Oscillations between predominantly clockwise and counter clockwise traveling waves were observed, which is clearly indicative of Bloch–oscillations [16].

In order to explain why the use of the Faraday rotator in the above–mentioned optical experiment is very promising for observing direct Bloch oscillations, it is instructive to see first what kind of problems arise if one would choose for the option with a *rotating* ring. As in the case of electrons, a crucial time scale is the coherence time, which is to be identified with the photon lifetime in the ring, the latter being limited mainly by (inevitable) scattering processes. Spreeuw et al. [16] have convincingly shown that it is very well possible to significantly enhance the photon life time in the presence of a population inverted gain medium in such a way that the roundtrip gain is just below the threshold for laser action[2]. In this way, a coherence time in the passive ring of the order of 0.1 ms is well within scope [18].

The other time scale involves the rate of change of the rotation frequency of the ring, $d\Omega/dt$, which leads to rather disencouraging conclusions. Namely, for one full Bloch–oscillation period we must create an extra phase 2π by varying the circulation within the photon lifetime over one fundamental quantum $C_0 = h/m_0$ (see (19)). This requires ($\Delta t = 0.1$ ms)

$$\Delta\Omega/\Delta t \gtrsim \frac{1}{2A}\frac{h}{m_0}\frac{1}{\Delta t} = \frac{9 \times 10^5}{A} \text{ rad/s}^2, \tag{23}$$

with the effective area A in square meters. So, one would have to apply enormous accelerations or use very large effective areas (e.g., a multi–loop ring resonator). In any case, it would not be easy to carry out the experiment in this way. Let us therefore now discuss the Faraday rotation alternative.

As the mere purpose of the rate of change of rotation frequency is to create an extra 2π roundtrip phase difference within the coherence time, this can also be accomplished by a rate of change of the ring's effective roundtrip length, e.g., by using a Faraday element. The Faraday rotator induces a phase difference between

---

[1]This inequality will be derived in the next section
[2]The theory here presented needs modification when it is to be applied in situations where net amplification of the waves takes place, such as in the case of ring laser operation (see also Chap.9 in this book).

counter propagating light waves with opposite circular polarizations. Using a magnetic induction (inside the Faraday element) of order 1 T, a phase difference of 2π can easily be realized. It can formally be shown [17] that, if used in combination with certain polarizing optical elements, the optical ring with Faraday element is indeed equivalent to a rotating ring. As the required rate of change of the magnetic induction $(dB/dt \simeq 10^4$ T/s) can be obtained without any special problems, it follows that the Faraday–rotation variant offers good prospects for observing the Bloch oscillation.

5. NON–ADIABATIC RESPONSE TO FAST PERTURBATIONS

In this section we will specifically deal with electrons, but in view of the analogy on the Schrödinger equation level many results and conclusions can directly be translated into the optical propagation domain. On the few occasions where this translation is not obvious, we will give the optical analogue special attention.

We now return to (9) with time–dependent flux $\Phi(t)$. It was recognized by Greenwood [19] in 1958 that a natural representation of the general solution of (9) is in terms of the $u_n(\Phi)$–functions introduced in (12). Indeed, for each value of Φ these functions form a complete and orthonormal set. So we write

$$\Psi(x,t) = \sum_n c_n(t) u_n(\Phi(t),x) \tag{24}$$

and try to express the behavior of the time–dependent coefficients $c_n(t)$ in terms of the $u_n(\Phi,x)$–eigenfunctions and the $E_n(\Phi)$–eigenenergy spectrum, where the latter are assumed to be known.

After substitution of (24) in (9), using (12) and assuming the spectrum $\{E_n(\Phi)\}$ nondegenerate for all Φ, we may arrive after a long and tedious derivation at the coupled equations for the coefficients $c_n(t)$ (given in Ref.[14]),

$$\frac{dc_n(t)}{dt} = \frac{-i}{\hbar} E_n(\Phi(t)) c_n(t) - \frac{i}{mL} \frac{e\hbar}{} \Phi(t) \sum_{l \neq n} \frac{D_{nl}(\Phi(t))}{E_{nl}(\Phi(t))} c_l(t), \tag{25}$$

where

$$D_{nl}(\Phi) \equiv \int_0^L dx \, \psi_n(\Phi,x)^* \frac{\partial}{\partial x} \psi_l(\Phi,x), \tag{26}$$

with ψ_n defined in (13) and

$$E_{nl}(\Phi) \equiv E_n(\Phi) - E_l(\Phi). \tag{27}$$

Eq.(25) allows for the following interpretation of the wave dynamics in the ring under the influence of the driving force $\Phi(t)$. There are two mechanisms which in general are coupled in a complicated manner. Firstly, there is this flow in reciprocal space (Φ–space here), causing the energy of a wave to run through the periodic energy bands (first term in (25)). If this were the only thing to happen we are describing the adiabatically slow dynamics discussed in the preceding section. However, there is this other mechanism described by the second term in the r.h.s.

of (25), by which transitions between energy bands may be induced, in a rate depending on the magnitude of Φ, thus mixing all waves in a complicated manner.

To the best of our knowledge, we were the first to recognize the existence of limiting situations for which the dynamics described by (25) may be solved [14]. If the potential $V(x)$ in (9) or (12) is sufficiently weak, the gaps in the energy spectrum $E_n(\Phi)$ are much smaller than the width of the energy bands. In that case the coupling described by the second term in (25) is only between two nearly degenerate bands, for which the denominator $E_{nl}(\Phi)$ becomes very small. This allows us to decouple (25) into successively (in time) independent 2×2 systems, so that the full time evolution can be obtained from the detailed analysis of one representative 2×2 system only. One such 2×2 system is sketched in Fig.2 for a case of electrons; the underlying parabolic E–Φ relation can still be recognized. In case of photons the underlying ω–Ω relation would be linear rather than parabolic.

Introducing new coefficients b, which are related to the c–coefficients in (25) through a diagonal unitary transformation (details can be found in Ref.14), each relevant 2×2 system leads to the coupled set of equations

$$\frac{db_+}{dz} = \frac{1}{2} \frac{\exp[i\gamma\int_0^z dy(1+y^2)^{\frac{1}{2}}]}{1 + z^2} b_- ,$$

$$\frac{db_-}{dz} = -\frac{1}{2} \frac{\exp[-i\gamma\int_0^z dy(1+y^2)^{\frac{1}{2}}]}{1 + z^2} b_+ .$$

$$(28)$$

where b_+ refers to the upper band and b_- to the lower band. Eq. (28) is to be integrated from $-z_0$ to $+z_0$; the dimensionless parameters z, z_0 and γ are given by

$$z = 2 \frac{\Delta E}{E_g} \frac{\Phi - \Phi_1}{\Phi_0}; \quad z_0 = \frac{1}{2}\frac{\Delta E}{E_g}; \quad \gamma = \frac{1}{2\hbar} \frac{E_g^2}{\Delta E} \frac{\Phi_0}{d\Phi/dt}.$$

$$(29)$$

Here, ΔE is the spacing between adjacent (unperturbed) energy levels (see Fig.2) and Φ_0 is the flux periodicity quantum $\Phi_0 = h/e$.

The coupled set of equations (28) can only give an adequate and representative description of the full dynamics if the coupling to other bands can be neglected. The following simple reasoning leads us to the conditions for which (28) is a valid approximation. The time dependence of Φ introduces a time dependence of the energy, $dE/dt = (d\Phi/dt)(dE/d\Phi) = (\Delta E/\Phi_0)d\Phi/dt$. In order to increase the energy by an amount E_1, say, we need a time interval $\delta t = E_1/(dE/dt)$. This interval is related to an uncertainty in the energy of at least

$$\delta E = \frac{\hbar}{2} \frac{\Delta E}{E_1} \frac{d\Phi/dt}{\Phi_0} .$$

$$(30)$$

This implies that bandstructure details smaller than this δE are meaningless for waves having acceleration $d\Phi/dt$. Hence, if we want to exclude transitions to adjacent bands separated by ΔE (see Fig.2) to occur, these bands should be well

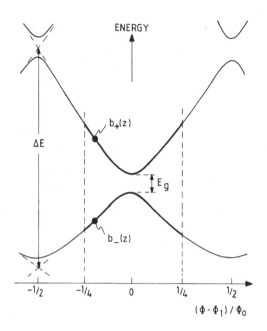

Fig.2   Illustrating the 2×2 system studied for Zener tunneling. Here, the flux Φ
is assumed to increase linearly in time, i.e., we move from left to right
along the horizontal axis. The coefficients $b_\pm$ are introduced in (28). Φ_1
is the flux value for which the two relevant unperturbed bands intersect
and Φ_0 is the flux periodicity quantum h/e.

outside each others acceleration–induced uncertainty δE. This implies (take $E_1 = \Delta E$)

$$\Delta E >> \delta E = \frac{\hbar}{2} \frac{d\Phi/dt}{\Phi_0}. \tag{31}$$

This condition was recently obtained by Lubin et al. [20]. It is independent of the
potential details, but it must be realized that the matrix element for a transition
to adjacent bands will tend to zero as the potential becomes weaker. For free
particles, there will be no transitions. So, the conditon (31) is sufficient but not
always necessary.
    Another consistency requirement for (28) to describe nontrivial dynamics is
that $E_g << \Delta E$. This condition follows if we demand transitions over the band gap
to occur without jumps to next–nearest bands; it reads $z_0 >> 1$, which implies a
weak scattering potential in general, while conditon (31) can be expressed as $\gamma >>
z_0^{-2}$.
    Eq.(28), or similar ones, have been extensively studied in the context of Zener
tunneling. We mention in this respect Zener [21], Stueckelberg [22], Houston [23],
Eilenberger [24], Lenstra et al. [19,25] and, recently, Mullen et al. [26]. The
equations are representative for a wealth of physical phenomena involving level

crossings and avoidings, such as in molecular and atomic collisions [27], Zener tunneling in semiconductors [23,28] and mode conversion phenomena between plasma waves [29].

Assuming the weak–potential condition $z_0 \gg 1$ satisfied, the value of the γ–parameter in (29) will crucially determine the nature of the solution of (28). This can be conveniently discussed in terms of the unitary matrix T given by

$$\begin{pmatrix} b_+(z_0) \\ b_-(z_0) \end{pmatrix} = \begin{pmatrix} T_{++} & T_{+-} \\ T_{-+} & T_{--} \end{pmatrix} \begin{pmatrix} b_+(-z_0) \\ b_-(-z_0) \end{pmatrix} . \tag{32}$$

Here, $|T_{-+}|^2 = |T_{+-}|^2$ is the probability for Zener tunneling, while $|T_{++}|^2 = |T_{--}|^2 = 1 - |T_{+-}|^2$ is the probability for a velocity–reversing process, i.e., a backscattering event. For $z_0 \gg 1$, (28) leads to [25]

$$|T_{+-}|^2 = \exp(-\tfrac{1}{2}\pi\gamma). \tag{33}$$

There are two extreme regimes according to (33), the case of typically slow dynamics, for which $\gamma \gg 1$, i.e., showing hardly any Zener tunneling at all, and the case of fast dynamics, i.e. $\gamma \ll 1$. In the latter case we have $|T_{--}|^2 = 1 - \exp(-\tfrac{1}{2}\pi\gamma) \simeq \tfrac{1}{2}\pi\gamma$. The absence of Zener tunneling in the first regime is compatible with the Bloch oscillation idea; it can be interpreted as a back and forth oscillation between backscattering events, or, to put it differently, between total Bragg reflections of the clockwise and counter clockwise waves by the periodic potential they experience.

We will now concentrate on the limit in which Zener tunneling is appreciable while there is only little back scattering. In this regime $\gamma \ll 1$ but in view of the consistency requirement (31) we also must have $\gamma \gg (z_0)^{-2}$. This means $(z_0)^{-2} \ll \gamma \ll 1$, which is not hard to satisfy if the scattering potential is sufficiently weak. Since the driving force $d\Phi/dt$ is large, the number of successive Zener–tunneling events per unit time will also be large. It takes two of these events when Φ varies over one flux quantum $\Phi_0 = h/e$, hence, there are $(e/\pi\hbar)(d\Phi/dt)$ Zener events per unit time. Now, suppose for the moment that these events are independent of each other. It would then make sense to introduce the backscattering rate

$$W = |T_{--}|^2 (e/\pi\hbar) \frac{d\Phi}{dt} \simeq \frac{E_g{}^2 L}{4\hbar^2 v} . \tag{34}$$

This result is precisely equal to what Fermi's golden rule of time–dependent perturbation theory would give us directly for scattering into a continuum of final states with level density $N(E) = 1/\Delta E = L/(2\pi\hbar v)$ and transition probability for weak scattering

$$\left| \frac{1}{L} \int_0^L dx \, e^{2ikx} \, V(x) \right|^2 = \tfrac{1}{4} E_g{}^2 .$$

Apparently, the dynamics of the waves is so fast that it effectively washes out the discrete level structure and turns it into a continuum. However, the above reasoning assumes independence of successive Zener events, which is, in fact, a wrong assumption. In the absence of any phase–disrupting process the waves are propagating strictly coherently implying successive Zener events to be connected by the *amplitudes* and not by the *probabilities*.

Various papers have been devoted to this problem of the time evolution in terms of successive coherently connected Zener events. This research was initiated by our 1986 paper [30], in which we reported a chaotic–type of time evolution of the electron wave function in the ring. We will not repeat the complete arguments here, but let us, before concluding this section, briefly sketch what is going on and discuss some of the latest developments. The chaotic evolution of the wave function can be traced back to the fact that, on accelleration, the dynamical phases evolve in an irregular, incommensurate way after each fundamental time period $\Phi_0/(d\Phi/dt)$. As a result of this, the electrical current in the ring as a function of time exhibits features that are expected for *diffusing* carriers much more than for coherently propagating waves: On time scales of the order of the elastic scattering time $1/W$, with W given by (35), the dynamics under the influence of a constant driving field, $d\Phi/dt=const.$, is very similar to what is predicted by a simple theory based on the Boltzmann equation with a collission term obtained from (35).

Due to close similarity with diffusively conducting electrons we deliberately used the adjective *resistive* to characterize the conduction behavior of electrons in the ring system [30]. Landauer [31] criticized this terminology by arguing that *resistive* must be reserved for a truly dissipative system whose behavior is irreversible in the time. The electron dynamics in the ring does have the time–reversal symmetry indeed, but we advocated the label resistive for a system whose current–field characteristics is resistive–like; i.e., linear [32]. This touches, in fact, a problem related to the second law of thermodynamics. Our Hamiltonian description of electronic conduction in a ring is in fact a pure–state description for which the entropy $S = \text{Tr}\, \rho \ln \rho$ vanishes and, hence, is constant in time. Of course, the strict constancy of entropy is intimately related to the lack of energy dissipation. However, an observer who is not informed about the underlying physics and who would base his entropy on the Boltzmann expression $\sum f_n \ln f_n$ with f_n occupation probabilities, such an observer would find increasing entropy in accordance with the second law. Clearly, this observer would see no difficulty in using the adjective *resistive*.

The resistive behavior only occurs during a restricted time interval, well exceeding $1/W$ but sufficiently short otherwise. This follows from similar calculations performed by others, in particular Blatter and Browne [33], who extended the time scale by two or more orders of magnitude and observed the current gradually decreasing down to zero. This was claimed to be caused by a sort of localization effect due to the coherent build up of the wave function in an increasing number of irregularly scattered waves. Gefen and Thouless [34] indicated how this time–dependent problem can be formally related to stationary coherent wave propagation in a 1–dimensional disordered system, for which wave localization (in space) is a well-known property. It should be realized that the dynamical localization effect can only be seen in a real ring if the electron coherence time can be made an order of magnitude larger at least than the elastic scattering time $1/W$, where the latter is of the order 10^{-9} s.

As a consequence of the above estimate, even the observation of elastic–scattering induced resistance in a ring may not be possible. However, it has been shown recently by Lenstra *et al.* [35] that the dynamical properties of electrons in a ring subject to a constant driving field $d\Phi/dt$ can be mapped onto the stationary properties of electrons in a tilted periodic lattice (the unfolded ring), where the tilting angle is proportional to $d\Phi/dt$. This tilting perturbs the periodicity of the lattice and causes elastic back scattering. The electrical resistance for a finite section of the lattice is via the Landauer formalism [36] directly related to transmission and reflection probabilities, where the latter can be determined by solving the time–independent Schrödinger equation. It was found [35] that electrons in the tilted lattice usually behave as if they were moving in a disordered potential, in which case localization is known to occur.

In view of modern semiconductor structure growth techniques by which Kronig–Penney–type superlattices of more than 100 unit cells, each cell consisting of 5–10 atomic layers, can easily be made, we expect that these systems are good candidates for observing the tilting–induced localization through measurements of the electrical resistance as a function of applied voltage (tilting angle). Such experiments when performed at low temperatures are very likely to increase our understanding of electrical resistance in the regime where elastic scattering is the dominant type of disturbance. The understanding from first principles of the residual, i.e. temperature $T \to 0$ K limit, resistance in a metal is a long–standing problem, closely related to the question as to whether at $T = 0$ K quantum–mechanical particles show diffusive behavior in the presence of elastic scattering objects (e.g., impurities, lattice imperfections).

Let us conclude this section by making a few remarks on an optical analogue of the tilted lattice. This situation can be realized by making a multilayer structure with the following dielectric function:

$$\epsilon_r(x) = \epsilon_r^{\text{periodic}}(x) + Fx, \tag{35}$$

where $\epsilon_r^{\text{periodic}}$ is a periodic function and Fx gives the tilting. If L is the periodicity length, the extra phase collected by a wave due to the tilting after n periods is

$$\varphi(n) = \tfrac{1}{2}FL^2(\omega_0/c)n^2. \tag{36}$$

The same effect on the phase would be obtained by increasing the length of the cells according to

$$L_n = L(1 + FLn^2), \tag{37}$$

which may be easier to manufacture. In any case, the phase angles $\{\tfrac{1}{2}FL^2(\omega_0/c)n^2 \bmod (2\pi)\}$ will be distributed pseudo–randomly in the interval $[0,2\pi]^3$, at least for not too small values of F, and we expect similar localization behavior to occur as for electrons in the tilted lattice. Unlike the electronic case, the optical variant would allow *direct* measurement of the reflection and transmission probabilities.

---

[3]Except for a set of F–values of measure zero related to very special discrete values of $\tfrac{1}{2}FL^2$.

ACKNOWLEDGMENT

This work is part of the research program of the Foundation for Fundamental Research on Matter (FOM), which is financially supported by the "Nederlandse Organisatie voor Wetenschappelijk Onderzoek" (NWO).

REFERENCES

[1]   Various contributions in: *Physics of Optical Ring Gyros*, S.F. Jacobs, M. Sargent III, M.O. Scully, J. Simpson, V. Sanders and J.E. Killpatrick, eds. Proc. SPIE 487 (1984).

[2]   P. Lett., W. Christian, S. Singh and L. Mandel, Phys. Rev. Lett. 47 (1981) 1892; L. Mandel, R. Roy and S. Singh, in *Optical Bistability*, eds. C.M. Bowden, M. Ciftan and Th. R. Robl (Plenum, New York, 1981).

[3]   W.R. Christian, T.H. Chyba, E.C. Gage and L. Mandel, Optics Commun. 66 (1988) 238.

[4]   R.J.C. Spreeuw, J.P. Woerdman and D. Lenstra, Phys. Rev. Lett. 61 (1988) 318.

[5]   C. Van Haesendonck and Y. Bruynseraede, Europhysics News 19 (1988) 89; R.A. Webb and S. Washburn, Physics Today 41 (Dec., 1988) 46.

[6]   T.I. Smith, Phys. Rev. Lett. 15 (1965) 460; J.E. Zimmerman and A.H. Silver, J. Appl. Phys. 39 (1968) 2679.

[7]   Many contributions on superconducting and normal–metal rings are found in: *SQUID '85*, eds. H.D. Hahlbohm and H. Lübbig (de Gruyter, Berlin, 1985).

[8]   R.A. Webb and S. Washburn, see Ref.[5]; see also B. Schwarzschild, Physics Today 41 (Jan., 1986) 17.

[9]   A short derivation is given in: D. Lenstra and W. van Haeringen, in: *Coherence and Quantum Optics 6*, eds. L. Mandel and E. Wolf (Plenum, New York, 1990).

[10] D. Dieks and G. Nienhuis, Am. J. of Physics, to be published.

[11] C. V. Heer, Phys. Rev. A134 (1964) 799; E.J. Post, Rev. Mod. Phys. 39 (1967) 475; also in Ref.[1].

[12] M. V. Berry, Proc. R. Soc. Lond. A392 (1984) 45; Scientific American (Dec., 1988) 26.

[13] B. Hendriks and G. Nienhuis, Quantum Optics, to be published.

[14] D. Lenstra and W. van Haeringen, J. Phys. C: Solid State Phys. 14 (1981) 5293.

[15] J.J. Haris, J.A. Pals and R. Woltjer, Review Repts. on Progr. in Phys. 52 (1989) 1217.

[16] R.J.C. Spreeuw, E.R. Eliel and J.P. Woerdman, *Bloch–Oscillations in the Photon Band Structure*, Optics Commun. (1990) in press.

[17] D. Lenstra and S.H.M. Geurten, *Optical Bandstructure and Polarization Modes in the Sagnac Ring Resonater*, Optics Commun. (1990) in press.

[18] R.J.C. Spreeuw, private communication.

[19] D.A. Greenwood, Proc. Phys. Soc. 71 (1958) 585.

[20] D. Lubin, Y. Gefen and I. Goldhirsch, *Mesoscopic Rings Driven by Time Dependent Magnetic Flux: Level Correlations and Localization in Energy Space*, preprint.

[21] C. Zener, Proc. Roy. Soc. Lond. A137 (1932) 696.

[22] E.G.C. Stueckelberg, Helv. Phys. Acta 5 (1932) 369.

[23] W. V. Houston, Phys. Rev. 57 (1940) 184.

[24] G. Eilenberger, Z. Phys. 164 (1961) 59.
[25] D. Lenstra, H. Ottevanger, W. van Haeringen and A.G. Tijhuis, Phys. Scripta 34 (1986) 438.
[26] K. Mullen, E. Ben–Jacob, Y. Gefen and Z. Schuss, Phys. Rev. Lett. 62 (1989) 2543.
[27] E.E. Nikitin and S.Ya. Umanskii, *Theory of Slow Atomic Collisions* (Springer–Verlag, Berlin, 1984).
[28] J.E. Ziman, *Principles of the Theory of Solids*, (Cambridge University Press, 1972) 190.
[29] D.G. Swanson, *Plasma Waves* (Academic Press, Boston, 1989) Sec. 6.3.
[30] D. Lenstra and W. van Haeringen, Phys. Rev. Lett. 57 (1986) 1623.
[31] R. Landauer, Phys. Rev. Lett. 58 (1987) 2150.
[32] D. Lenstra and W. van Haeringen, Phys. Rev. Lett. 58 (1987) 2151.
[33] G. Blatter and D.A. Browne, Phys. Rev. B 37 (1988) 3856.
[34] Y. Gefen and D.J. Thouless, Phys. Rev. Lett. 59 (1987) 1752.
[35] D. Lenstra, W. van Haeringen and R.T.M. Smokers, Physica A (1990) in press.
[36] R. Landauer, Philos Mag. 21 (1970) 863; this volume Chap.15.

D. Lenstra and *W. van Haeringen* are with the Department of Physics, Eindhoven University of Technology, P.O. Box 513, 5600 MB Eindhoven, The Netherlands; *D. Lenstra* is also with the Huygens Laboratory, University of Leiden, 2300 RA Leiden, The Netherlands.

WHAT TO EXPECT FROM SIMILARITIES BETWEEN THE SCHRöDINGER AND MAXWELL EQUATIONS

M. Kaveh

We discuss interference phenomena in random systems for electrons and for optical waves and the similarities between the Schrödinger and Maxwell equations. In particular, we describe the following phenomena: (i) weak localization, (ii) coherent backscattering peak, (iii) memory effect, (iv) fluctuations and their statistics, and (v) correlations. In each case, we compare the effects for electrons and for photons and present the available experimental evidence.

1. INTRODUCTION

The subject of the propagation of waves in random media was developed almost independently for electron waves (obeying the Schrödinger equation) [1] and for electromagnetic waves (obeying the Maxwell equations) [2]. The recent breakthrough [3,4] in our understanding of electron transport phenomena in random systems stemmed from the realization that electrons are quantum waves and as such, interfere. The new idea is that the phase of the scattered wave is "remembered" after a sequence of elastic scattering events caused by the randomness of the media. This is reproducible and a wave that undergoes the same sequence of elastic scattering events will always acquire the same phase. This means that interference phenomena must play an important role even in random media and that the propagation of the wave is not purely diffusive. The interference phenomena slow down the diffusive propagation of the wave and result in the phenomenon of "weak localization". This concept was developed by the condensed matter physicists [3,4]. However, similar interference phenomena were developed for optical waves with a different terminology [5–7]. Only very recently have the similarities between optical and electron waves been realized [8,9].

The purpose of this chapter is to discuss these similarities (and differences), to point out the generality of certain interference phenomena and to discuss some experiments which test these ideas. An electron wave carries a charge and a spin. An optical wave, on the other hand, is a vectorial wave. This by itself leads to many different phenomena. In particular, the effect of electron–electron interactions is of great importance [4] in understanding the transport properties of disordered materials. This is a very complex subject which is not yet well established for strong disorder. The validity of a single electron wave equation in the case of a many–electron system is always questionable. Nevertheless, when the disorder is weak (namely, when the transport elastic mean free path ℓ is much larger than the wavelength λ), many experiments and much theoretical work lead to the conclusion that the basic interference phenomena (discussed below) can be adequately described by a single–electron wave equation. For stronger disorder

21

W. van Haeringen and D. Lenstra (eds.), Analogies in Optics and Micro Electronics, 21–34.

($\ell \simeq \lambda$), one needs a many–electron wave equation and the similarity between the Maxwell equations and the Schrödinger equation breaks down. Only for non–degenerate electron systems (where electron–electron interactions are extremely weak) will the analogy hold. The non–universality in the critical behavior of transport exponents [10] is believed to result from electron–electron interactions [11]. Thus, the nature of the Anderson localization transition, which is a pure disorder transition, is hoped to be experimentally resolved mainly by optical waves.

2. THE WAVE EQUATION

The general scalar wave equation is given by

$$\{\nabla^2 + K^2[1 + \mu(r)]\}\psi_\omega(r) = 0, \tag{1}$$

where $\psi_\omega(r)$ can describe either the wave function of an electron or a component of an electromagnetic field. For electrons, $K^2 = 2mE/\hbar^2$, and $\mu(r) = -V(r)/E$, where m is the mass of the particle, E is its energy and $V(r)$ is the potential. For electromagnetic waves, $\mu(r)$ is the fluctuating part of the refractive index and $K = \omega/c$ where c is the average speed of light in the medium. Eq. (1) is an approximation both for electrons and for optical waves. For electrons, it neglects electron–electron interactions (or treats them within some one–electron effective potential) and for optical waves it neglects their vector nature. As will be pointed out later these effects lead to specific different predictions and expectations. Most of our discussion will deal with phenomena which result from (1). Some of these phenomena may sometimes apply better to optical waves rather than to electron waves and sometime vice versa, due to the *approximations* involved in converting the Maxwell equations and the many–electron Schrödinger equation to (1).

Even (1) has never been solved exactly. Due to the complexity of the random potential, $\psi_\omega(r)$ will fluctuate from one sample to another for different realizations of $\mu(r)$. Thus, averaged quantities are required. Indeed, the most common approach which has been adopted in optics [2] and in solid state physics [12] is to write an averaged transport equation for $<|\psi(r,t|^2> \equiv n(r,t)$, where $< >$ denotes an ensemble average and $\psi(r,t)$ is the time–dependent wave solution. It can be shown that, to good approximation, the averaged transport equation can be converted into a diffusion equation

$$D_0\nabla^2 n(r,t) = \partial n(r,t)/\partial t. \tag{2}$$

This means that propagation of charge or energy as given by (1) is diffusive and the distance of propagation is given by random–walk theory,

$$<|r-r_0|^2> = \int n(r,t)(r-r_0)^2 d^3r = D_0 t, \tag{3}$$

where D_0 is the diffusion constant given by $D_0 = (1/d)v\ell$, where d is the dimensionality of the system, and v is the velocity of the wave.

The conversion of (1) into (2) was believed to be a good approximation for weak disorder where $\ell \rangle \lambda$. Only after 1979, was it realized [3] that this mapping is in principle incorrect. Eq. (1) contains the time–reversal property for *each* realization of $\mu(r)$. Thus, $n(r,t)$ must also obey an equation with time–reversal symmetry which is *not* the case for the diffusion equation given by (2). It is this very point which leads to new interference phenomena which result from (1) but are absent from (2). Eq. (2) is basically a *particle* equation, since $< |\psi(r,t)|^2 >$ neglects the *phases* of the wave. The phases of the wave are important even in random systems and lead to unique interference phenomena. Before discussing these effects, it should be pointed out that the most dramatic prediction regarding the propagation of waves in random systems was made by Anderson [13] and developed extensively by Mott [1]. This is the concept of localization. Anderson predicted in 1958 that when the fluctuations in the electron static potential in (1) become large enough, the wave ceases to diffuse and becomes localized. This follows from (1) from the fact that the stationary wave solutions of (1), for strong disorder, are confined in space and fall off from some point r_0 in the system as $\exp[-|r-r_0|/\xi]$ with a localization length ξ which depends on the electron energy. Mott suggested that the Anderson transition will occur when $\ell \simeq \lambda/2$, and that at this critical disorder, the diffusion constant D_0 will vanish. Most of the earlier (before 1979) treatments for calculating D_0 were based on perturbation theory which showed that (2) is a good approximation of (1).

The Kubo formulation [1], which accounts for the wave properties of the electrons, seemed to indicate that the wave nature of the electrons for weak disorder is not important. Thus, the Anderson transition was viewed as a transition from particle–like behavior described by (2) to wave–like behavior described by (1), which results in localization. In this picture, the Anderson transition must be of first order, namely, the diffusion constant D_0 drops to zero discontinuously. This conclusion is valid as long as (2) is valid down to the transition. For then Mott argued that D_0 can never be zero, but retains its minimum value when ℓ reaches its minimum value of $\lambda/2$. In this case, $D_{min}=v\lambda/6$. Mott's picture of the Anderson transition was modified upon finding new wave properties which follow from (1). This leads to a diffusion constant D which is smaller than D_0 which is obtained from the particle picture described by (2)

3. WEAK LOCALIZATION

Using perturbation theory to solve (1), it was found that certain diagrams ("maximally crossed") contribute significantly to D. The interpretation of these diagrams in k–space [4] or in real–space [14] made it clear why D is always smaller than D_0 and may drop *continuously* to zero [3,15]. The reduction of D is termed "weak localization". The understanding of weak localization is very simple in terms of the interference of waves. Any wave which undergoes multiple elastic scattering and travels from point A to point B in the system can arrive by many different Feynman trajectories. The probability $P_{AB}\equiv P$ for a wave to arrive at B (after starting from A) is given by

$$P = |\sum_i A_i|^2 = \sum_i |A_i|^2 + \sum_{i \ne j} |A_i||A_j|\cos(\phi_i-\phi_j) \tag{4}$$

where A_i is the probability amplitude for a specific Feynman trajectory. The first term in (4) is just the sum of individual probabilities to arrive at B. This is the random walk probability which is particle–like, since it does not contain the phases of the wave. We denote the first term by $P_0 = \sum |A_i|^2$. The second sum in (4) results from interference between different wave trajectories and, as we shall see later, this sum is responsible for many fluctuation phenomena discovered recently

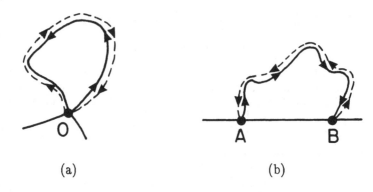

Fig.1   (a) Time–reversed trajectories which interfere at point O.
         (b) Time–reversed trajectories which lead to the CBP.

in transport or optical measurements. At this stage we are interested in D, which is usually an ensemble averaged property. The point is that when P is ensemble averaged, the second term in (4) vanishes since ϕ_i and ϕ_j are assumed to be random independent variables. Thus, $<P> = P_0$ and no correction from interference of waves enters $<P>$. However, this is not the case when point B is made to coincide with point A. In this case, P is the probability of returning to the origin. Now, two different Feynman trajectories will contribute to $<P>$ because for each trajectory that starts at A and ends at A, there is a "time–reversed" trajectory as shown in Fig.1a. These conjugated trajectories exist primarily because (1) obeys time–reversal symmetry. The phases of each pair of time–reversed trajectories are the *same* and equal to $(2\pi/\lambda)W_j$, where W_j is the length of the specific trajectory. This unique time–reversal wave property increases P by a factor of 2, leading to $P = 2P_0$. The interference between the two time–reversed trajectories is independent of the specific potential realization and therefore survives the ensemble average, yielding $<P> = 2P_0$. Thus, the wave probability to return to the origin is twice as large as the particle probability. This leads to a reduction in the diffusion constant which depends on dimensionality. The dimensionality dependence of the reduction comes from a phase–space argument [4,14–16] taking into account the fact that the interference is still effective if point B is within a wave–length of point A. This leads to [16]

$$D/D_0 = \begin{cases} 1-(3/(2\pi)^2)(\lambda/\ell)^2[1-\ell/L_0] & ; \qquad d = 3 \\ 1-(1/2\pi^2)(\lambda/\ell)\ln(L_0/\ell) & ; \qquad d = 2 \end{cases} \qquad (5)$$

where L_0 is the *smallest* trajectory length for which time–reversal symmetry is preserved. Thus, D for a wave depends on a length scale L_0. Eq. (5) is universal [15]. In Fig.2, we present the first confirmation [17] of the universal length dependence of D in two–dimensional systems where L_0 results either from a magnetic field or from inelastic scattering which destroys the time–reversal symmetry. Applying the same ideas to the transmission of optical waves, we get $T = (\ell/L)(D/D_0)$, where T is the averaged transmission coefficient and D/D_0 is given by (5). Although (5) was widely observed for electrons [4], this correction to

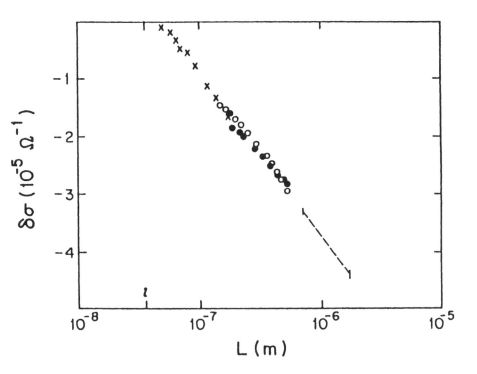

Fig.2   The change in the conductivity vs. the log of the shortest length in the
          system for different magnetic fields, temperatures and electric fields.
          (after Kaveh et al. [17]).

the transmission coefficient of optical waves has not yet been observed.

It was argued [15] that for $d = 3$, (5) can be extrapolated down to the
transition $\ell \simeq \lambda$ which leads to a critical exponent $\nu=1$, namely $D=D_0(\lambda-\lambda_c)/\lambda_c$,
for $L_0 = \infty$, and to $D = D_0(\ell/L)$ for finite systems, in agreement with the
one–parameter scaling theory [3]. The scaling theory thus predicts that the
diffusion constant of a wave in a three–dimensional random system drops
continuously to zero. At the transition, the diffusion is anomalous and
scale–dependent. Using the scaling relation $D = D_0(\ell/L)$ near the transition leads
to [9] $T = (\ell/L)^2$, a prediction which has not yet been confirmed experimentally.

We now turn to an important implication of weak localization on backscattered
light from a random system. Suppose that the closed trajectory is now opened as in
Fig. 1b. Any wave that travels from A to B interferes with a wave that travels in
the time–reversed direction from B to A. If the *initial* phases at A and B are the
same, one gets constructive interference. The phases at A and B can be made the
same if these points are chosen near the surface of the incident wave. The
backscattered wave will have a peak enhanced exactly by a factor of 2 in the
backscattered direction, where k_f points opposite to the incident direction k_i,
namely at $k_f = - k_i$. The factor 2 arises from the same time–reversal symmetry as

in weak localization. This phenomenon is called the enhanced coherent backscattered peak (CBP) which was recently observed by several groups in random fluids [18,19] (suspensions of polysterene spheres in water) and in random solids [20,21]. Such a CPB is shown in Fig. 3a for an amorphous material. It is interesting that the first prediction and correct quantitative diagrammatic calculation of the CBP was made by Barabarenkov [5] back in 1973. Unfortunately, this paper was not known to the solid state physics community which discovered the importance of the maximally crossed diagrams only in 1979. The CBP was calculated for radio waves and no connection was made to weak localization.

The recent awareness of the CBP is due mainly to the first two clear observations in fluids and later to the two observations in solids and to the successful theoretical treatment of its detailed shape by Akkermans and Maynard [22]. It was also pointed out [21] that the fact that the CBP was observed at all was due to the rapid motion of the scatterers which ensemble average the interference between different Feynman trajectories thus forming a fluctuating speckle pattern. Moreover, the CBP in random solids is hidden and one usually observes a speckle pattern as in Fig.3b. Kaveh et al. [21] suggested that rotating the sample is equivalent to ensemble averaging and indeed the speckle of Fig.3b disappeared and a nice CBP was revealed (see Fig.3a). The general expression for $I(q)$, where $q=k_i+k_f$, is given by [21]

$$I(q) = I_0\{1+ \sum_{\ell>m} |P_{\ell m}|^2\cos[q\cdot(r_l-r_m)]+C(k_i,k_f)\}, \qquad (6)$$

where $|P_{1m}|^2$ is the probability that the wave trajectory that started at r_l will end up at r_m and $C(k_i,k_f)$ is a four–fold sum [21] which depends on the wave phases. The first sum leads to the CBP and the second sum is the fluctuating speckle. The ensemble averge removes the second sum. The CBP exhibits the same form for a strictly two–dimensional system in the weak–disorder lmit (where $L << \xi$). The CBP was recently observed for a two–dimensional system [23]. We thus see the similarity between a weak–localization phenomenon and the CBP effect. In the back–scattered direction, $q = 0$, and $<I(q)> = 2I_0$. Thus, this leads to an enhancement by a factor of 2, exactly as for the case of weak localization where $<P> = 2P_0$ according to (4). We may use this similarity to study time–reversal phase–breaking mechanisms in the CBP in the same manner as in transport measurements.

The CBP will depend on various length scales L_0 in a similar way as the diffusion constant D in (5). However, for optical waves the main length scales are the finite size of the sample L and the absorption length $L_a=(D\tau_a)^{1/2}$ (where τ_a is the time in which the wave is absorbed). These length scales tend to round off the triangular shape of the CBP around scattered angles of order $\Theta \simeq \lambda/L_0$. Thus, as in (5) the CBP can be used to extract the different phase breaking length scales L_0. One may also apply this idea of CPB of *electrons*. For example, it was recently suggested by Berkovits et al. [24] that one measure the CBP of electrons in the presence of a magnetic field. In this case, the form of $I(q)$ was calculated exactly for a given temperature where $I(q)$ depends both on L_i (the inelastic length scale) and on the magnetic length scale $L_H=(ch/eH)^{1/2}$. It was shown that to a good approximation $I(q,L_H,L_i)$ can be extracted from $I(q,L_H=\infty,L_i=\infty)$ (The CPB in the absence of any length scale) by replacing q with $(q^2+(L_i)^{-2}+L^{-2})^{1/2}$. This effective length dependence of $I(q)$ is similar to the effective length dependence

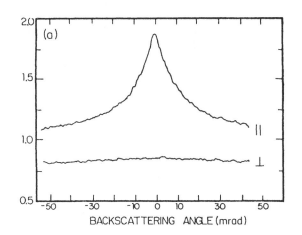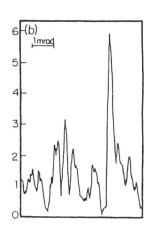

BACKSCATTERING ANGLE (mrad)

Fig.3   (a) Coherent backscattering peak for a rotating sample of BaSo$_4$ with
           parallel ($\parallel$) and perpendicular ($\perp$) configurations [21].
           (b) Intensity fluctuations due to the second term in (6) [21].

which was first suggested by Kaveh *et al.* [17] in the study of universal length scale
dependencies of the conductivity as demonstrated in Fig.2.

Another interesting phase–breaking mechanism for electrons is the spin–orbit
interaction with a new length scale L_{so}. For trajectories smaller than L_{so}, one gets
the usual weak–localization effect due to constructive interference. However, if the
trajectory in Fig. 1a is larger than L_{so}, this leads to the concept of antilocalization
which results from *destructive* interference. Antilocalization has been widely
observed [25] in the study of magnetoresistance of materials with strong spin–orbit
interaction.

It was recently suggested by Berkovits and Kaveh [26] that the role of the
different Feynman trajectories in the presence of spin–orbit coupling can be
directly visualized by studying the CBP. It was shown that for angles $\Theta < \lambda / L_{so}$, an
*anti*enhancement should be observed. This results from the fact that long
trajectories contribute to the interference of CBP only for small angles. In Fig.4 we
plot the CBP of electrons in the case of strong spin–orbit interactions. We see that
we get a separation of the contributions of different Feynman trajectories. The
large Feynman trajectories contribute to a dip in the CBP due to destructive
interference. Thus, for $\Theta < \lambda / L_{so}$. we get an antienhancement. This prediction [26]
has not yet been confirmed experimentally.

What about antilocalization of optical waves? Assuming scalar waves, there is
no analogous situation for optical waves. However, the role of the electron spin is
replaced by the polarization. If one sends an incident wave polarized in the x
direction and looks for the CBP for the emitted wave polarized in the y direction,
theory predicts [18,19,27] and experiments confirm [18,19,21] a very small CBP
due to destructive interference. Nevertheless, theory always predicts an
enhancement. It was, however, demonstrated by Freund [28] for the case of three
mirrors specifically arranged that one gets an antilocalization effect due to the
vector nature of the wave. Whether a real material exists with such a property is a
challenging question.

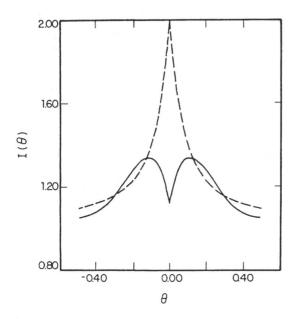

Fig.4   Backscattered intensity as a function of backscattered angle. The solid
            curve corresponds to the case of spin–orbit scattering and the dashed
            curve is the CBP without spin–orbit scattering [26].

4. MEMORY EFFECT

We have seen that the main development in our understanding of transport
properties at very low temperatures is the concept of quantum interference. This is
an interference effect which results from the time–reversal symmetry of (1). The
fact that electrons interfere like waves is based on the idea that the phase memory
is *never* lost due to multiple elastic scattering. An interesting and quite surprising
phenomenon, termed the "memory effect", was recently suggested [29,30] and
confirmed experimentally [31]. The idea was to test whether a wave transmitted
through a random slab of tickness L "remembers" the initial wave direction k_i, of
the incident wave. The prediction is that an observer of the transmitted wave can
determine the change in the (unseen) incident wave direction k_i. It turns out that
the speckle pattern at a given direction k_f *follows* the direction of k_i for small
changes in angle $\Theta < \lambda/L$. The confirmation of this important prediction means
that elastic multiple scattering of a wave is *not* a Markovian process. Each
scattering event remembers the *previous* scattering event and the phase is
accumulated deterministically. This important conclusion results from the fact
that any incident wave determines uniquely the wave solution of (1) in the random
medium. This wave solution acts like an agent for transfering the information
contained in the incident wave. The incident wavefront selects the particular wave

solution in the medium and acts as a boundary condition. Any small change in the initial wave front will result in a small change in the wave solution and consequently a change in the wave front of the transmitted wave. A real–space theory of the effect was given by Berkovits et al. [31]. This theory is based on Feynman trajectories and provides a convincing explanation of the memory effect. The memory effect is thus the basis of the concept of wave interference due to multiple elastic scattering.

5.   FLUCTUATIONS AND THEIR STATISTICAL PROPERTIES

We now return to (4) which gives the probability P for a wave to arrive at some point B after starting from point A. Certainly, P will fluctuate from sample to sample due to different realizations of the positions of the scatterers. From (4), we can easily calculate all the moments of P to get $<P^n> = n!<P>^n$. This simply results from the fact that we may have $n!$ possibilities of arranging $2n$ indices to be equal in order to obtain non–zero ensemble averaged quantities. This leads to the result $<I^n> = n!<I>^n$ where I is the intensity of the wave at point B. This is a well–known result first obtained by Lord Rayleigh. It means that the intensity at a given point will fluctuate from sample to sample with an exponential distribution function $P(I) = <I>^{-1}\exp[-I/<I>]$. This result was also obtained diagrammatically [32] by using the ladder diagrams. Thus, Rayleigh statistics is based on interferences between different, *independent*, Feynman trajectories as given in (4).

We now discuss the connection between the fluctuations in electron and optical systems. One of the most important predictions in this field is termed "universal conductance fluctuations" (UCF) which states that the conductance of a disordered system fluctuates from sample to sample in such a way that its variance $(<G^2>-<G>^2)^{1/2}$ is universal. Namely,

$$\text{Var } G = e^2/h. \tag{7}$$

This was proven diagrammatically [33]. The analogy between optical waves and electrons that results from (1) suggests that a simpler proof of (7) may exist. The main insight to the UCF is that it results from the fact that the different Feynman path trajectories that contribute to the conductance are statistically *dependent*. Suppose they were independent. We may then easily calculate the full distribution of G by using the Landauer formulation in which

$$G = e^2/h \sum_{\alpha\beta} T_{\alpha\beta}, \tag{8}$$

where $T_{\alpha\beta}$ are transmission coefficients with N incident channels α (which are the different k_i directions with the same ($|k_i|$) and N outgoing channels β. The number of channels is $N=(k_1)^2A=4\pi^2A/\lambda^2$, where A is the area of slab surface. We may define the total transmission due to the N incident channels as $\sum_{\alpha\beta} T_{\alpha\beta}=T$. The UCF is then general for any wave and results in Var $T = 1$. For optical waves we usually have one incident wave front k_i (a plane wave) and the transmission coefficient is then given by $T_\alpha = \sum_\beta T_{\alpha\beta}$. Stephen and Cwilich [34] calculated diagrammatically the second moment of T_α, obtaining

$$<(T_\alpha)^2> - <T_\alpha>^2 = \alpha(\lambda^2/A)(\ell/L), \tag{9}$$

where $\alpha = 9/2\pi$. This result means that the different $T_{\alpha\beta}$ in (9) are correlated. These correlations result from long range spatial correlations.

To see this more clearly, let us suppose that the outgoing intensities $T_{\alpha\beta}$ are independent and obey exponential statistics $P(x) = <x>^{-1}\exp[-x/<x>]$ where $x = T_{\alpha\beta}$. Adding N independent $T_{\alpha\beta}$ terms leads to

$$P(T_\alpha) = \{N^N T^{N-1}/((N-1)! <T_\alpha>^N)\}\exp[-NT_\alpha/<T_\alpha>], \qquad (10)$$

which yields for the second moment

$$<(T_\alpha)^2> - <T_\alpha>^2 = <T_\alpha>^2/N. \qquad (11)$$

Using $<T_\alpha> = \ell/L$, we get

$$<(T_\alpha)^2> - <T_\alpha>^2 = (\ell/L)^2/N, \qquad (12)$$

which is smaller by a factor of (ℓ/L) than the result obtained by Stephen [34] as given by (9). This means that the $T_{\alpha\beta}$ are correlated. A simple argument to obtain (9) follows from an idea of Imry [35] about the effective number of channels. The incident wave contains N spatial channels whereas the transmission must contain only $N(\ell/L)$ channels. Thus, the effective number of transmitted channels is $N_{\text{eff}} = N(\ell/L)$. This means that we have only N_{eff} uncorrelated new quantities $T_{\alpha\beta}$ and N_{eff} must replace N in (11). This yields, instead of (12), the result

$$<(T_\alpha)^2> - <T_\alpha>^2 = (\ell/L)/N, \qquad (13)$$

which is in agreement with the diagrammatic calculation (up to a numerical factor).

We see that even if the incident channel α is kept fixed (as in a transmission experiment for an optical wave), we get larger fluctuations than would be expected from uncorrelated Feynman trajectories.

We now turn to a simple argument [36] to get UCF based on optical wave statistics [37]. We define a total reflection coefficient $R = \sum_{\alpha\beta} R_{\alpha\beta}$, similar to the total transmission coefficient T. Conservation of energy requires

$$R + T = N, \qquad (14)$$

which leads to Var $T =$ Var R. Assuming that the $R_{\alpha\beta}$ are independent, we get equations similar to (10,11) but with T replacing T_α and N^2 replacing N (since we have N^2 terms to sum). Thus,

$$<R^2> - <R>^2 = <R^2>/N^2. \qquad (15)$$

For $\ell/L << 1$, we get from (14) $<R> \simeq N$, which when inserted in (15) leads to Var $R = 1$ and to UCF. This argument assumed that the $R_{\alpha\beta}$ are independent which is not exactly the case [38]. Nevertheless, the above argument demonstrates that UCF are *not* sensitive to these correlations.

We see that the fluctuations decrease systematically as we go from $T_{\alpha\beta}$ to T_α and then to T. The commulant $<x^2> - <x>^2$ is equal to $<x>^m$ where $m = 2$ for $x = T_{\alpha\beta}$, $m = 1$ for $x = T_\alpha$ and $m = 0$ (universal) for $x = T$. It should be noted that although the long–range correlations tend to increase the *relative* fluctuation $(<x^2>-<x>^2)/<x>^2$ for T_α and T, it remains a very small quantity. It is equal to N^{-1}, where N is the number of effective channels. For $x = T_{\alpha\beta}$, $N=1$ and it is equal to unity and decreases dramatically for T_α and further for T. UCF was confirmed by measuring [39] the magnetic–field dependence of the conductance for "mesoscopic" systems where $L_i < L$. The fluctuations in T_α have not yet been observed, but were confirmed numerically [40].

We have seen that the advantage of optical waves is that we can study *intermediate* situations. The conductance is a multichannel transmission T whereas we are now able to study the one incident channel transmission T_α and even the intensity at a given point $T_{\alpha\beta}$. In the weak–disorder limit, it is believed that the distribution functions for these quantities form a single–parameter scaling function. Namely, for $T_{\alpha\beta}\equiv x$, we have $P(x) = <x>^{-1}\exp(-x/<x>)$, and for T_α (or T), $P(y)=(2\pi)^{-1/2}\exp(-y^2/2)$ where $y\equiv(T_\alpha-<T_\alpha>)/\mathrm{Var}\ T_\alpha$. In Fig.5, we plot the numerical results of Edrei *et al.* [40] of $P(T_\alpha)$ for a two–dimensional disordered system (in the weak–disorder limit, $\ell>>\lambda$) which confirms the Gaussian form. For strong disorder it was predicted [41] that for T and T_α, the distribution function turns into a log–normal distribution. Recent calculations claim [42] that the distribution functions for T or T_α cannot be described by a single–scaling parameter function over the entire range of disorder. For strong disorder, there is a need for at least two parameters to describe $P(y)$. This raises serious doubts about the hypothesis of a single–parameter scaling theory [3] for the Anderson transition. Thus, the nature of the Anderson transition is still an open problem. The optical Anderson transition is very rich in unique interference phenomena of which we here discussed only the weak–disordered limit. In particular, the following effects near the transition are of great interest. The pulse shape of transmitted waves, the CBP, the memory effect, intensity correlations and statistical properties of the fluctuations. Recently, detailed predictions of these phenomena were made [43] based on the scaling theory. These predictions may help future experiments to examine to what extent a single–parameter scaling theory is valid.

6. CORRELATIONS

It was shown by Thouless [44] that the relevant energy scale for electrons in the weak–disorder limit is $E_\eta=\hbar D/L^2$. Indeed, this is the energy scale which causes a significant phase change for an averaged Feynman trajectory of N steps in a slab. The argument is as follows [45]. The averaged phase acquired by the wave after N multiple elastic scattering events is $\phi = (2\pi/\lambda)N$. The change in the phase due to a small change $\Delta\lambda$ in the wavelength is $\Delta\phi = (2\pi/\lambda)N(\Delta\lambda/\lambda)$. Thus, the phase of a given multiple Feynman trajectory is extremely sensitive to small changes $\Delta\lambda$, since the total change in ϕ is multiplied by the entire length of the trajectory. Longer trajectories are therefore more sensitive to $\Delta\lambda$. Equating $\Delta\Phi$ to $\pi/4$, we get that the dephasing occurs for $\Delta\lambda/\lambda = \lambda/(8N)$. In the weak–disorder limit, for a slab geometry, $<N> = (\ell/L)^2$. Converting into energy changes by using $E=(2\pi/\lambda)\hbar v$, yields the Thouless result $\Delta E=\hbar D/L^2$. Taking into account the specific dephasing of all different trajectories, one can calculate the intensity correlation function $C(\Delta\lambda) = <I(\lambda)I(\lambda+\Delta\lambda)>-<I(\lambda)>^2$. It was found that this energy correlation function scales as $C(L^2\Delta E)$. This correlation function was calculated within the factorization approximation in which $C(\Delta\lambda) = |<E^*(\lambda)E(\lambda+\lambda)>|^2$. This approximation was checked numerically [45] and was found to be excellent. Moreover, it is also in excellent agreement with recent

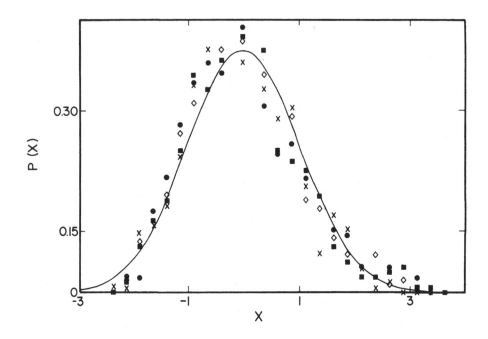

Fig.5   Numerical distribution function [40] as a function of $\chi =$
          $\left(T_\alpha - <T_\alpha>\right)/\text{Var } T_\alpha$. The solid curve represents a Gaussian function
          $(2\pi)^{-1/2}\exp(-x^2/2)$.

experiments [46].

One can calculate the energy correlation functions for T_α and T either
diagrammatically [29,34] or by using a Langevin approach [47]. Here too the
Thouless energy plays an important role. However, for $\Delta E >> \hbar D/L^2$, it was shown
[34] that the factorization approximation fails and the long–range nature of the
spatial correlations introduces long tails into the correlation function $C(\Delta E)$. For
such large changes in E, it is expected that T_α and T will fall off as a power law
rather than exponentially as is the case for $\Delta E < \hbar D/L^2$. The long–range behavior
of $C(\Delta E)$ for T_α or T has not yet been observed.

7. SUMMARY

We have seen that converting the Maxwell equations as well as the many–electron
Schrödinger equation into a unified wave equation leads to similar wave
phenomena both for electrons and for optical waves. The weak–disorder limit now
seems to be quite well understood, containing many exciting interference
phenomena. The next step is to confine electrons into narrow constrictions forcing
them to behave as electron wave guides. This new area has begun and beautiful
physics has already emerged [48]. The solutions of the Maxwell equations in wave
guides will turn out to be of utmost importance in understanding the transport

properties of electrons under geometrical confinement.

As regards the Anderson localization transition, much work has yet to be done both theoretically and experimentally. For optical waves it turns out [49] that strong disorder is also often followed by strong absorption, which is a real obstacle to a definitive study of the nature of the Anderson transition. On the other hand, the electron systems are complicated by the fact that electron–electron interactions are important, in particular near the transition, and a simple wave equation as in (1) is not valid.

ACKNOWLEDGMENTS

The author acknowledges important collaboration with Sir Nevill Mott and M. Pepper on electron transport phenomena in disordered systems and with R. Berkovits, I. Edrei, I. Freund, M. Rosenbluh and B. Shapiro on localization phenomena in optical waves.
This work was supported by the Basic Research Administration of the Israel Academy of Science and Humanities and by the Israel–US Binational Science Foundation.

REFERENCES

[1]   N.F. Mott and E.A. Davis, *Electonic Processes in Non–Crystalline Materials* (Oxford University Press, 1979).
[2]   A. Ishimaru, *Wave Propagation and Scattering in Random Media* (New York: Academic, 1978).
[3]   E. Abrahams, P.W. Anderson, D.C. Licciardello and T.V. Ramakrishnan, Phys. Rev. Lett. 42 (1979) 673.
[4]   For reviews, see G. Bergmann, Phys. Rep. 107 (1984) 1; P.A. Lee and T.V. Ramakrishnan, Rev. Mod. Phys. 57 (1985) 287; N.F. Mott and M. Kaveh, Advan. Phys. 34 (1985) 329.
[5]   Yu. N. Barabarenkov, Izv. Vyssh. Ucheb. zav. Radiofizika 16 (1973) 88.
[6]   Y. Kuga and A. Ishimaru, J. Opt. Soc. Am. A1 (1984) 831.
[7]   Y. Kuga and A. Ishimaru, J. Opt. Soc. Am. A1 (1984) 836.
[8]   S. John, Phys. Rev. Lett. 53 (1984) 2169.
[9]   P.W. Anderson, Phil. Mag. B52 (1985) 502.
[10]   G. T homas, Phil. Mag. B50 (1984) 1169 and references therein.
[11]   M. Kaveh, Phil. Mag. Lett B52 (1985) L1.
[12]   J. Ziman, *Electrons and Phonons* (Clarendon Press, Oxford, 1960.)
[13]   P.W. Anderson, Phys. Rev. 109 (1958) 1492.
[14]   D.E. Khmelnitskii. Physica B126 (1984) 235.
[15]   M. Kaveh and N.F. Mott, J. Phys. C15 (1982) L707.
[16]   M. Kaveh, Phil. Mag. B56 (1987) 693.
[17]   M. Kaveh, M.J. Uren, R.A. Davies and M. Pepper, J. Phys. C14 (1981) L413.
[18]   M.P. Van Albada and A. Lagendijk, Phys. Rev. Lett. 55 (1985) 2692.
[19]   P.E. Wolf and G. Maret, Phys. Rev. Lett. 55 (1985) 2696.
[20]   S. Etemad, R. Thompson and M.J. Andrejco, Phys. Rev. Lett. 57 (1986) 575.
[21]   M. Kaveh, M. Rosenbluh, I. Edrei and I. Freund, Phys. Rev. Lett. 57 (1986) 2049.
[22]   E. Akkermans and R. Maynard, J. Phys. Lett. 46 (1985) L1045; E. Akkermans, P.E. Wolf and R. Maynard, Phys. Rev. Lett. 56 (1986)1471.

[23] I. Freund, M. Rosenbluh, R. Berkovits and M. Kaveh, Phys. Rev. Lett. 61
      (1988) 1214; R. Berkovits and M. Kaveh, J. Phys. C20 (1987) L181.
[24] R. Berkovits, D. Eliyahu and M. Kaveh, Phys. Rev. B (to be published).
[25] G. Bergmann, Solid State Commun. 42 (1982) 815.
[26] R. Berkovits and M. Kaveh, Phys. Rev. B37 (1988) 584.
[27] M.J. Stephen and G. Cwilich, Phys. Rev. B34 (1986) 7564.
[28] I. Freund, Phys. Rev. A37 (1988) 1007.
[29] S. Feng, C. Kane, P.A. Lee and A.D. Stone, Phys. Rev. Lett. 61 (1988) 834.
[30] R. Berkovits, M. Kaveh and S. Feng, Phys. Rev. B40 (1989) 737.
[31] I. Freund, M. Rosenbluh and S. Feng, Phys. Rev. Lett. 61 (1988) 2328.
[32] B. Shapiro, Phys. Rev. Lett. 57 (1986) 2168.
[33] P.A. Lee and A.D. Stone, Phys. Rev. Lett. 55 (1985) 1622; B.L. Altshuler
      and D.E. Khmeltnitskii, Pis'ma Zh Eksp. theo. Fix. 42 (1985) 291 (JETP
      Lett. 42 (1986) 359).
[34] M.J. Stephen and G. Cwilichs, Phys. Rev. Lett. 59 (1987) 285.
[35] Y. Imry, Europhys. Lett. 1 (1986) 249.
[36] P.A. Lee, Physica 140A (1986) 169.
[37] J.W. Goodman, J.Opt. Soc. Am. 66 (1976) 1145.
[38] P.A. Mello, E. Akkermans and B. Shapiro, Phys. Rev. Lett. 61 (1988) 459.
[39] S. Washburn and R.A. Webb, Advan. Phys. 35 (1986) 375.
[40] I. Edrei, M. Kaveh and B. Shapiro, Phys. Rev. Lett. 62 (1989) 2120.
[41] P. Mello, J. Math. Phys. 271 (1986) 2876; B. Shapiro, Phil. Mag. 56 (1987)
      1031.
[42] B.L. Altshuler, V.E. Kravtsov and I. Lerner, JETP Lett. 43 (1986) 44w1.
[43] R. Berkovits and M. Kaveh, J. Phys. C (to be published).
[44] D.J. Thouless, Phys. Rev. Lett. 39 (1977) 1167.
[45] I. Edrei and M. Kaveh, Phys. Rev. B38 (1988) 950; *ibid*, Phys. Rev. B (to be
      published).
[46] A. Genack, Phys. Rev. Lett. 58 (1987) 2043.
[47] B.Z. Spivak and A. Yu Zyuzin, Solid State Commun. 65 (1988) 311; R. Pnini
      and B. Shapiro, Phys. Rev. B39 (1989) 6986.
[48] M. Pepper, Phys. World, 1 October (1988) 45.
[49] A. Genack and J. Drake (preprint).

M. Kaveh is with Bar–Ilan University, Department of Physics, Ramat–Gan 52100,
Israel.

SYNERGETICS AS A THEORY OF
ANALOGOUS BEHAVIOR OF SYSTEMS

H. Haken

This article deals with nonlinear devices in optics and micro–electronics and unearths the roots of their analogous behavior. At the microscopic level, elementary excitations such as photons, electrons, holes, Cooper–pairs, polaritons, etc. may be produced. When their number increases, cooperative nonlinear effects occur. In most cases of practical interest, the high excitation level permits a classical treatment of the originally quantum–mechanical processes by means of quantum–classical correspondence. In the vicinity of instability points (e.g. onset of oscillations) the order parameter concept and the slaving principle allow us to represent the dynamics by one or a few order parameters which can be grouped into universality classes describing analogous behavior. An example of how these results can be applied to pattern recognition is given.

1. INTRODUCTION

In this article I wish to show how synergetics may help us to unearth analogies between devices in optics and micro–electronics. In this way a deeper understanding of the properties of those devices can be gained and in addition principles for the construction of new devices may be developed. Synergetics is an interdisciplinary field of research that I initiated some twenty years ago. It deals with systems composed of many individual elements or parts which may produce spatial, temporal or functional structures by self–organization [1–3]. The question that has been asked by synergetics from the very beginning was whether there are general principles which govern the formation of those macroscopic structures irrespective of the nature of subsystems. As it turned out, such principles could be found indeed, provided we focus our attention on situations in which the macroscopic behavior of a system changes qualitatively. A typical example for such qualitative changes on a macroscopic scale is provided by the laser [4]. When a laser is pumped only weakly, it acts as a lamp and emits individual wave tracks so that microscopic chaos of light is produced. When the pump power exceeds a critical value, a well–ordered coherent light wave with a macroscopic amplitude is produced, however. Though the laser is a system driven far from equilibrium, the transition from its disordered to its ordered state shows a remarkable formal analogy to phase transitions of systems in thermal equilibrium, such as in superconductors or ferromagnets, as was shown by Graham and Haken [5] as well as by DeGiorgio and Scully [6].

In this article I shall treat a still far broader class of analogies which hold between active devices. In this context it is a good starting point to distinguish between linear and nonlinear devices. In optics, lenses, mirrors, and gratings, as

W. van Haeringen and D. Lenstra (eds.), Analogies in Optics and Micro Electronics, 35–47.
© 1990 *Kluwer Academic Publishers. Printed in the Netherlands.*

well as optical wave guides are linear devices as are Ohmic conductors or resistors
in electronics. We shall not discuss these devices here. On the other hand there is
an increasingly important class of devices based on nonlinear processes. These are
in optics in particular lasers, optical parametric amplifiers and nonlinear crystals
leading to frequency mixing, optical rectification and so on. In micro–electronics
examples are provided by, e.g., tunnel diodes, Gunn oscillators and Josephson
junctions.

As we shall demonstrate below, the analogies between these devices become
apparent at the "order parameter level" where the macroscopic behavior of these
devices can be described by one or a few characteristic variables which show
analogous behavior. A good deal of processes in optics and in micro–electronics is
concerned with information, namely its storage, its transfer, and its processing by
logical elements. This leads to requirements for the devices, such as reliability,
speed and compactness, and also for the need to have rapid access to them.
Information storage can be achieved by steady states of bistable devices, e.g.
tunnel diodes, or of multi–stable devices. It may also be realized by oscillations.
Information transfer may be achieved, in particular, by electrons, holes, photons or
phonons; information processing, for instance by parametric oscillators. The
elements can then be implemented in highly integrated circuits. As we shall see
below, synergetics allows us to think also of still more highly integrated devices in
which the logical elements are in a way continuously distributed over the system.

2. AT WHICH LEVEL DO THE ANALOGIES BECOME VISIBLE?

Both in optics as well as in micro–electronics, the starting point for our
consideration may be elementary excitations. In optics these are the photons while
in solid state physics they may be provided by electrons, holes, excitons, phonons,
polarons, polaritons, Cooper–pairs of superconductors, spinwaves and may be other
kinds of excitations. While some of these elementary excitations are always
present, such as conduction electrons in metals, in many cases they are produced
by various processes. Photons are produced by light emission from excited atoms
or semiconductors, for instance by light emitting diodes, conduction electrons,
holes, or excitons in semiconductors may be produced by absorption of light and so
on. At this stage, the production rate of elementary excitations is proportional to
the input energy. But then internal processes may set in, such as impact
ionization, eventually leading to dielectric breakdown, or oscillations or
filamentation in semiconductors, stimulated emission of photons leading to an
exponential growth of their number and so on. Eventually, nonlinear effects
become important which lead to a saturation or to specific wave–wave interactions
as in optical parametric oscillators.

At this stage, at least in general, the elementary excitations reach a
macroscopic size. The electric field strength of the laser or an optical parametric
oscillator behaves as a classical coherent field on which only small quantum
fluctuations of phase and amplitude are superimposed. The electric current of the
Gunn oscillator or the phase of the Josephson junction become macroscopically
measurable quantities etc. The macroscopically measurable quantities, or in other
words, the information carrying quantities, are the order parameters. Examples for
them are provided by the laser light field strength or the amplitude of the current
(or its lowest Fourier component) of the Gunn oscillator. What is crucial in the
present context is the fact that the order parameters obey specific classes of
equations. Order parameters of the same class behave dynamically the same way.

This allows us to realize the same macroscopic behavior by physically (or chemically or even biologically) quite different devices. Which device to be chosen will depend on the specific task, e.g. time constants, reliability, signal to noise ratio etc. It will be beyond the scope of this article to discuss all the classes known so far so that we provide the reader with some typical examples (which characterize most of the known devices, however).

1) Bistable devices may be described by

$$\frac{d\xi}{dt} \equiv \dot{\xi} = \alpha + \lambda\xi + \gamma\xi^2 - \delta\xi^3 + F(t), \tag{1}$$

where α, λ, γ, δ are system parameters, ξ the (time–dependent) order parameter, and $F(t)$ a stochastic force representing internal or external noise. By a suitable linear transformation of ξ, α or γ can be made equal zero. A special case of (1) is

$$\dot{\xi} = \lambda\xi - \delta\xi^3 + F(t),$$

which we discuss in somewhat more detail. To visualize the behavior of the order parameter ξ, we interpret (purely formally) ξ as the coordinate of a particle with mass m and add the corresponding acceleration term to the l.h.s. of (1)

$$m\ddot{\xi} + \dot{\xi} = \lambda\xi - \gamma\xi^3 + F(t) . \tag{2}$$

Eq. (2) is the equation of a particle moving under the force $\lambda\xi - \delta\xi^3$, which may be derived from a potential V, and under the fluctuating force $F(t)$,

$$m\ddot{\xi} + \dot{\xi} = -\frac{\partial V}{\partial \xi} + F(t) . \tag{3}$$

For $m \to 0$, we recover (1), but we have now a simple visualization of the behavior of ξ at hand, viz. the overdamped motion of a particle in the potential of Fig.1.

Quite clearly, bistability is present in Fig.1b, so that the corresponding device can store one bit of information. Landauer [7] compared the switching from one stable state (of a tunnel diode) to the other one with the shift of a Bloch–wall in a ferromagnet, while Graham and Haken [5] and DeGiorgio and Scully [6] compared the system's behavior (of a laser) with that of a ferromagnetic phase transition when λ changes sign, i.e. when Fig.1a is replaced by Fig.1b. Phenomena such as symmetry breaking, critical slowing down, and critical fluctuations occur in the transition region $\lambda \approx 0$.

2) Nonlinear oscillations (single mode laser, parametric oscillator, Gunn oscillator, etc.) close to threshold. The order parameter ξ is complex,

$$\dot{\xi} = (-i\omega + G) \xi - C|\xi|^2 \xi + F(t) , \tag{4}$$

where ω is the frequency, G the gain, C the saturation constant, and $F(t)$ the

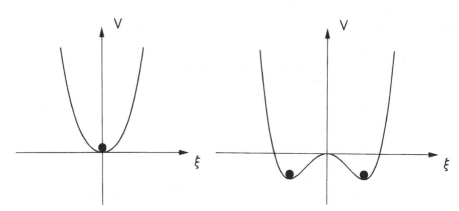

Fig.1   Visualization of the time–variation of an order parameter; (a) for $\lambda < 0$,
             one attractor at $\xi = 0$; (b) for $\lambda > 0$, two attractors at $\xi \neq 0$.

fluctuation force. Again the time dependence of ξ can be visualized as that of a
particle in a potential.

3) Phase–locking between the two phases ϕ_1, ϕ_2 of two oscillators, where the
relative phase $\phi_2 - \phi_1 = \phi$ is the order parameter

$$\dot{\phi} = K - A \sin \phi + F(t) \, . \tag{5}$$

K, A are system parameters.

4) Multistability; the order parameter q is a vector with components q_1, q_2,.., q_n
which obeys the equation

$$\dot{q} = - \, grad_q \, V(q) + F(t) \, . \tag{6}$$

The only fixed points are stable or unstable nodes, or saddles. We shall encounter
an important example in Sec. 5.

3.   FROM ELEMENTARY EXCITATIONS TO MACROSCOPIC BEHAVIOR;
AN OUTLINE OF THE THEORY

In this section we give an outline of the theoretical concepts and tools that are
needed to proceed from the elementary excitations to the order parameters and
their equations that provide us with an adequate macroscopic description of the
device under consideration. We start with the description of elementary
excitations. Phonons and photons are treated as bosons with their creation and
annihilation operators $b_k^\dagger$, b_k, respectively. Within a certain approximation
excitons, Cooper pairs and magnons can also be described as bosons. On the other
hand, electrons and holes must be described as fermions with their creation and
annihilation operators $a_k^\dagger$, a_k, respectively. Because fermions are, at least in the

context of materials, conserved quantities, the creation of a particle in state k by an operator $a_k{}^\dagger$ is connected with the annihilation of that particle in the same or in a different state j. Therefore, in the theories, the operators appear pairwise in the form $a_k{}^\dagger a_j$. Cooper pairs are described by operators of the form $a_k{}^\dagger a_j{}^\dagger$, for the creation of such a pair and by $a_j a_k$ for their annihilation.

In the first step of the analysis, the Hamiltonian must be established. This Hamiltonian for the individual excitations reads in the case of Bose–operators

$$\sum \hbar\omega_k \, b_k{}^\dagger b_k \tag{7}$$

and in the case of Fermi–operators

$$\sum E_k \, a_k{}^\dagger a_k \, , \tag{8}$$

where $\hbar\omega_k$ and E_k are the energies of the state labelled by index k. The interaction Hamiltonian acquires different forms depending on the nature of the particles, i.e. whether they are bosons or fermions or Cooper pairs. In the case of the interaction between fermions, the typical Hamiltonian consists of contributions of the form

$$a_k{}^\dagger a_j a_l{}^\dagger a_m \, , \tag{9}$$

while the interaction between fermions with bosons acquires the form

$$a_k{}^\dagger a_j \, b_m{}^\dagger, \; a_k{}^\dagger a_j \, b_l. \tag{10}$$

In this case the transition of a fermion from one state to another one is connected with a generation or annihilation of a boson. The interaction between Cooper pairs consists of elements of the form

$$a_k{}^\dagger a_j{}^\dagger a_l a_m \, . \tag{11}$$

In nonlinear optics in a number of cases the interaction between the individual waves can be described by polynomials of $b^\dagger$ and b, e.g. in the form

$$b_k{}^+ b_j{}^\dagger b_l \, . \tag{12}$$

In all the cases discussed, the total Hamiltonian H_0 of the system consists of a superposition of the terms described in $(7) - (12)$. In general, such a Hamiltonian is not sufficient for the description of a device, however. A device interacts with its surrounding, for instance by receiving energy from it or dissipating energy into it. This interaction is taken care of by a general Hamiltonian H_{Bath} and the interaction Hamiltonian H_{int}, describing the interaction between the bath and the system under consideration. The total Hamiltonian thus reads

$$H = H_0 + H_{\text{Bath}} + H_{\text{int}}. \cdot \tag{13}$$

Going into the details of the further treatment of the system by means of (13) is far beyond this present article. We, therefore, just wish to indicate briefly the main steps. Two methods have been particularly useful, namely (for a survey cf. [8]) 1. the Heisenberg equations of motion and 2. the density matrix equation.

In the case of the Heisenberg equations of motion, the individual operators undergo a time evolution while the state vector is time–independent. Denoting one of the corresponding operators $b_j{}^\dagger b_k$ or $a_j{}^\dagger a_k$ by Ω, the corresponding Heisenberg equation of motion reads

$$\dot{\Omega} = \frac{i}{\hbar} [H, \Omega], \tag{14}$$

where the bracket denotes the usual commutation operation. Denoting the density matrix by ρ, the density matrix equation is given by

$$\dot{\rho} = -\frac{i}{\hbar} [H, \rho] . \tag{15}$$

In both cases (14) and (15), the bath variables can be eliminated and give rise to additional terms in (14) and (15), leading to equations of the form

$$\dot{\Omega} = \frac{i}{\hbar} [H', \Omega] + \Gamma\Omega \tag{16}$$

and

$$\dot{\rho} = -\frac{i}{\hbar} [H', \rho] + \Gamma'\rho . \tag{17}$$

The additional terms $\Gamma\Omega$ or $\Gamma'\rho$ describe the damping and fluctuations caused by the interaction of the proper system described by H_0 with bath variables. When we take adequate expectation values of (14), the equations can be cast into classical equations which can be easily interpreted as equations for classical amplitudes. A still more elegant method is provided by the principle of quantum–classical correspondence (cf. [8, 9]). In this case a connection is established between the quantum–mechanical density matrix and a classical distribution function $f(q)$ by means of an equation of the form

$$f(q) = \int \exp[-iq\beta] \, \mathrm{tr}(\exp(i\beta\Omega)\rho) \, d^N\beta . \tag{18}$$

There are some problems how to define the exponential function of the set of operators Ω because in general the operators do not commute. We refer the reader for a treatment of this problem to the literature [8,9]. We mention only that one may show that $f(q)$ obeys a generalized Fokker–Planck equation of the type

$$\dot{f}(q,t) = Lf . \tag{19}$$

L is a linear operator which differentiates f with respect to the classical variables q. When the size of elementary excitations becomes macroscopic, in many cases the generalized Fokker–Planck equation can be replaced by an ordinary Fokker–Planck equation to which the corresponding Langevin equation (or Ito or Stratonovich equation) can be established [10,11]. This equation can be written in the form

$$\dot{q} = N(q, \alpha) + F, \tag{20}$$

where N is a nonlinear function of the state vector q, α represents a set of control parameters which describe external signals or pumping acting on the system, e.g. the energy input into a laser, and F represents internal and external fluctuating forces which are always present when damping or pumping is taken into account in a system.

At this stage we may apply some general ideas of synergetics [1,2]. We assume that for a certain range of a control parameter α_0 a solution of q_0 is known. In many cases this may be a well-known time- and space-independent state. When the control parameter α_0 is changed, this old solution

$$\alpha_0 : q_0 \tag{21}$$

may become unstable. In order to check the stability, the hypothesis

$$\alpha_0 \rightarrow \alpha \quad q = q_0 + w \tag{22}$$

is made and inserted into (20). As long as fluctuations are neglected and only linear terms are kept, (20) transforms into

$$\dot{w} = L\, w. \tag{23}$$

The general solutions of (23) can be written in the form

$$w_j = \exp(\lambda_j t) v_j , \tag{24}$$

where in the absence of degeneracies v_j is time-independent and in the other case may contain a finite number of powers of t. We introduce the vectors v_j as a new basis and make the hypothesis

$$q = q_0 + \sum \xi_j(t)\, v_j . \tag{25}$$

For what follows we shall distinguish between the unstable modes where

$$j = u, \ \mathrm{Re}\, \lambda_j \geq 0, \tag{26}$$

and the stable modes where

$$j = s, \ \mathrm{Re}\, \lambda_j < 0. \tag{27}$$

Inserting (25) into (20) and projecting the resulting equations on the eigenvectors v_j, we obtain the equations

$$\dot{\xi}_u = \lambda_u\, \xi_u + N_u(\xi_u, \xi_s) + F_u(t) , \tag{28}$$

and

$$\dot{\xi}_s = \lambda_s\, \xi_s + N_s(\xi_u, \xi_s) + F_s(t) , \tag{29}$$

where we made the distinction (26) and (27).

As can be shown for all devices so far known which use an instability, only very few or only one variables ξ_u appear, while most of the variables belong to the class of stable modes. As may be shown by the rigorous slaving principle of synergetics [1,2], the stable modes ξ_s can be expressed by the unstable modes ξ_u

$$\xi_s = f_s(\xi_u, t) . \tag{30}$$

This means that we may eliminate the amplitudes ξ_s in (28) so that via the resulting equation

$$\dot{\xi}_u = \lambda_u \, \xi_u + \tilde{N}_u(\xi_u) + F_u(t) \tag{31}$$

and (30) the dynamical behavior of the system is entirely determined. The very important consequence of this result is the following:

The dynamics of a system (device) close to its "instability point" (where coherent macroscopic behavior sets in) is governed by only one or a few variables, ξ_u, that are called the order parameters. Once they are determined by means of the solution of equations (31), the behavior of all the other variables ξ_s of the system is determined. The time–dependence in (30) stems from the fluctuations F_s, F_u only so that when we neglect this time–dependence, ξ_s is a uniquely determined function of ξ_u. These results have several far reaching consequences. Namely in many cases close to the instability point, the order parameters ξ_u are small quantities. This allows us to expand N_u into a power series of ξ_u and to keep only the leading terms, in general up to second or third order. The resulting equations acquire then a very simple form, e.g.

$$\dot{\xi} = \lambda \, \xi - C \, \xi \, |\xi|^2 + F(t), \tag{32}$$

in the case of a single complex order parameter. Thus we obtain the equations discussed in Sec.2 (since we treat only a single variable, we have omitted the index u).

4. WHAT USE CAN BE MADE OF THE ANALOGIES?

Above we have given a brief outline of a fairly abstract procedure which starts from the Hamiltonian describing the interaction between elementary excitations among themselves and with their surrounding. Then we found classical equations for amplitudes which, in a way, can be interpreted as amplitudes of waves, such as light waves, or density waves of electrons, and so on. Eventually it turned out that the behavior of the total system can be described by means of equations referring to one or a few order parameters. These order parameters describe a specific macroscopic physical state. Because the fluctuations are in such a case much smaller than the size of the order parameter, the system can reliably store information. Also transitions between the states exhibited in Fig.1 can be caused, e.g. by lowering the potential barrier between the two minima or by feeding energy into one of the states.

In this case information can be processed and logical elements realized. The essential message from our results is the following: Optical devices as well as

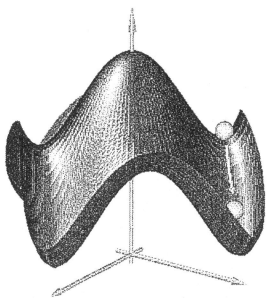

Fig.2   Example of the dynamics in a potential V.

devices of micro–electronics may lead to the same order parameter equation. The order parameters stand for the information to be stored or processed, or, in other words, information processing can be performed by the same order parameters and their equations which in turn can be realized by a variety of different processes and substrates. Thus in order to achieve a specific function of a device, we have to solve, in a way, an inverse problem. Instead of going from microscopic processes to the macroscopic level, we start here from the macroscopic level describing the functioning of the device. Then we describe this level by the adequate order parameter equations and eventually have to seek microscopic processes which give rise to the required macroscopic process. The recipe to perform this task is to look through the order parameter equations for various devices and see whether the required order parameter behavior is realizable by one of the known microscopic mechanisms. This is certainly not a simple task but in the remainder of this article I wish to show how this can be achieved in a most timely problem, namely the realization of neural computers. Since our type of neural computer is based on principles of synergetics, I have called it "synergetic computer". In addition it will turn out that it is basically different from the traditional neural computers what the kind of elements is concerned.

5. SYNERGETIC COMPUTERS FOR PATTERN RECOGNITION

In this section I wish to show how the general methodology described above can be applied to construct physical models of a computer for parallel computation and for the recognition of patterns [12]. I hope in this way I can show quite explicity how our approach works in a concrete case and may stimulate similar procedures in other cases. To perform pattern recognition, we shall use three ingredients:

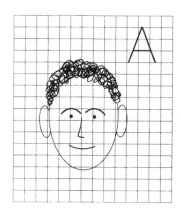

Fig. 3   Digitalization of patterns.

a) The associative memory concept (cf. [13]). Once a set of data is given, the system must be able to complement the set of data in a definite manner. For instance a telephone book is an associative memory because once we know the name and first name, the telephone book provides us with the telephone number. Similarly when we know the face and its name and the face is shown to us, we wish to remember the name.

b) The system is assumed to work as a dynamic system in which the state vector describing the system moves in some potential landscape, e.g. that of Fig.2. The minima correspond to the stored patterns, places away from the minima correspond to incomplete or distorted or noisy patterns. The dynamics has to restore the pattern or does away the noise by pulling the system into the corresponding attractor state to which the initial state has been nearest.

c) The property of a synergetic system based on the slaving principle. When dynamic structures are formed, initially a part of the system may be in an ordered state. This part then generates its specific order parameter which in turn forces the whole system into the ordered state. In analogy to this process, in associative memory a set of features is given. These features generate their order parameter which then provides the system with the lacking features. To demonstrate explicitly how our approach works, we proceed as follows: We first introduce a number of prototype patterns in the form of faces with their names which are labelled by letters A, B, etc.. The photographs are then digitized, e.g. by 60 by 60 pixels, the grey value of each pixel represents the component of a vector v (Fig.3). The individual vectors are distinguished by an index u. We now construct a dynamics by which a test pattern vector q is transformed in $q(t)$ which eventually approaches one of the test pattern vectors v_u to which it has been closest. As one may show in detail, this dynamics can be realized by the following equations [12]

Fig. 4   Stored prototype patterns (faces and letters encoding the names).

$$\dot{q} = \sum \lambda_u \, v_u(v_u{}^+q)$$
$$- B \sum_{u \neq v} (v_v{}^+q)^2 \, (v_u{}^+q) \, v_u$$
$$- C \sum_{u,v} (v_v{}^+q)^2 \, (v_u{}^+q) \, v_u \, . \tag{33}$$

Here the vectors $v_u{}^+$ are the adjoint vectors obeying the equations

$$(v_u{}^+v_v) = \delta_{uv} \equiv \begin{cases} 1 \text{ for } u = v \\ 0 \text{ for } u \neq v \end{cases} \tag{34}$$

In order to introduce the order parameter concept, we multiply (33) by $v_u{}^+$ and introduce the order parameters by means of

$$\xi_u = (v_u{}^+ \, q). \tag{35}$$

The equations of the order parameters then read

$$\dot{\xi}_u = \xi_u(\lambda_u - B \sum_{v \neq u} \xi_v^2 - C \sum_v \xi_v^2). \tag{36}$$

So here we are at the abstract level in which we describe the pattern recognition process by means of order parameters. It can be shown that the order parameter ξ_j approaches 1 provided q was closest to v_j and to zero otherwise. We solved the equations (36) on a serial computer and calculated

$$q(t) = \sum_u \xi_u(t) \, v_u \, . \tag{37}$$

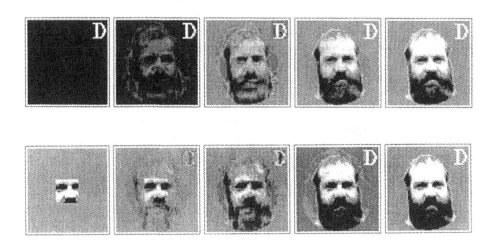

Fig. 5   From left to right: completion of originally presented patterns.

Fig. 4 represents some of the stored prototype patterns, while Fig. 5 shows the
results of the solution of (36) and (37) for specific test patterns (first frame on the
l.h.s.) (after Fuchs and Haken [14]).

Now it is our purpose to realize the order parameter equations by means of a
physical system. Such a system is provided by a fluid heated from below. It is well
known that such a fluid may form rolls which can be oriented in different
directions so that the system shows multistability. The order parameter equations
of the rolls are identical with the equations (36) provided a few idealizations are
made, which preserve the qualitative behavior of the solution, however. Fig. 6
shows computer calculations which exhibit the multistability. Thus our example
shows quite explicitly that multistability needed for pattern recognition can be
realized by the microscopic processes occurring in fluids. One may expect that the
order parameter equations (36) can be realized by suitable micro–electronic
elements or by lasers. By means of the analogy at the order parameter level a new
device can be constructed. In the present case its construction consists of two
steps: one step in which the projection

$$q_j \rightarrow \xi_u \tag{38}$$

is done via (35), which was not discussed here, and the second step in which
multistability is exploited. Quite evidently, in the case of a fluid there are no
logical elements, such as in the by now conventional neural computers, but
nevertheless multistability is achieved. Of course, for practical applications our
example of a fluid is not quite adequate because a fluid forms its structures at a
rather slow time rate. In practice, the step (38) may be done by holography,
whereas specific arrangements of lasers which share a common region in which gain
saturation takes place, possess order parameter equations of the form (36).

Fig. 6   Example of multistability in fluids.

REFERENCES

[1]   H. Haken, *Synergetics, An Introduction. Nonequilibrium Phase Transitions and Self–Organization in Physics, Chemistry and Biology* (Springer, Berlin, Heidelberg, New York, 1983).
[2]   H. Haken, *Advanced Synergetics. Instability Hierarchies of Self–Organizing Systems and Devices* (Springer, Berlin, Heidelberg, New York, 1987).
[3]   H. Haken, *Information and Self–Organization, A Macroscopic Approach to Complex Systems* (Springer, Berlin, Heidelberg, New York, 1988).
[4]   H. Haken, *Light II: Laser Light Dynamics* (North–Holland Elsevier, Amsterdam, New York, 1985).
[5]   R. Graham and H. Haken, Z. Physik 213 (1968) 420; 237 (1970) 31.
[6]   V. DeGiorgio and M.O. Scully, Phys. Rev. A2 (1970) 117a.
[7]   R. Landauer, IBM J. Res. Dev. 5 (1961) 3; J. Appl. Phys. 33 (1962) 2209; Ferroelectrics 2 (1971) 47.
[8]   H. Haken, Rev. Mod. Phys. 47 (1975) 67.
[9]   H. Haken, *Laser Theory* (Springer, Berlin, Heidelberg, New York, 1983), Chap.IX.
[10]   R.L. Stratonovich, *Topics in the Theory of Random Noise,* Vol. 1 (Gordon and Breach, New York, London, 1963).
[11]   C.W. Gardiner, *Handbook of Stochastics Methods* (Springer, Berlin, Heidelberg, New York, 1983).
[12]   H. Haken, *Neural and Synergetics Computers*, ed. H. Haken (Springer, Berlin, Heidelberg, New York, 1988).
[13]   T. Kohonen, *Self–Organization and Associative Memory* (Springer, Berlin, 1987).
[14]   A. Fuchs and H. Haken, *Pattern Recognition and Associative Memory as Dynamical Processes in a Synergetic System I+II*, Erratum; Biol. Cybern. 60 (1988) 17–22, 107–109, 476.

H. Haken is with the Institut für Theoretische Physik und Synergetik der Universität Stuttgart, Pfaffenwaldring 57/4, 7000 Stuttgart 80 (Vaihingen), F.R.G.

DENSITY OF STATES
OF ELECTRONS AND ELECTROMAGNETIC WAVES
IN ONE–DIMENSIONAL RANDOM MEDIA

Paul Erdös and Zbigniew Domanski

The knowledge of the phase $\theta(x)$ of the quantum–mechanical wavefunction $\psi(x)$ of an electron in a one–dimensional random potential $V(x)$ can be used to determine the density of states as a function of electron energy. A class of random potentials, called multi–step potentials, is introduced. A set of differential equations is formulated for the probability density of the phase. The equations are solved for two– and three–step potentials, and the results are discussed. In the discussion an analogy with a jogger (or rotator) model is helpful. The density of states is calculated numerically. The results may also be used to discuss the eigenstates of electromagnetic waves in layered media.

1. INTRODUCTION

The question, what are the quantum–mechanically allowed energy states of a particle in a one–dimensional random potential, was first studied numerically in 1954 [1] and then analytically in 1960 [2]. Subsequent work showed, that this question has to be formulated much more specifically, because the answer depends crucially on the type of randomness of the potential. We will forgo here a classification of random potentials, and the corresponding allowed energy states, and refer to published reviews [3,4].

There is a close relationship between the behavior of an electron in a one–dimensional random potential, and the behavior of the electromagnetic field in a medium with random dielectric constant. The mathematical treatment of both phenomena leads to the same type of equations.

This chapter is concerned with a class of random potentials for which general results for the density of states were previously not available. This class of potentials, defined in Sec. 3, seems to be particularly close to real physics – which justifies its interest. It also has the merit, that a very striking analogy with a rotator model can be developed, which allows visualisation of the phenomena involved.

In Secs. 2–5 the method of stochastic differential equations will be described for the calculation of the density of states. The rotator analogy, called the joggers' marathon, is developed in Sec. 6. The Secs. 7 and 8 contain the application of the method to the above–mentioned class of potentials, but it should be pointed out, that the method is not restricted to this class.

Particularly simple special cases are treated in Secs. 7 and 8. Ref. [5] contains a first short report on this work. Results for a special case of the random potential

49

W. van Haeringen and D. Lenstra (eds.), Analogies in Optics and Micro Electronics, 49–68.

here studied were obtained in [6]. A discussion follows in Sec. 9.

2. THE AVERAGE NUMBER OF STATES

Consider the Schrödinger equation for a particle of mass m moving in the interval $(0,L)$ under the influence of an arbitrary time–independent potential $U(x)$, subject to the usual regularity requirements,

$$-\frac{\hbar^2}{2m}\psi''(x)+[U(x)-E]\psi(x) = 0, \qquad \psi'' \equiv \frac{d^2\psi}{dx^2}, \qquad \psi(0) = \psi(L) = 0. \quad (1)$$

We shall make use of Sturm's oscillation theorem [7] concerning boundary–value problems associated with second–order linear differential equations. According to this theorem a real solution $\psi(x)$ of such an equation belonging to the eigenvalue E_m has m zeros in the interval $(0,L)$ of the variable x. This theorem applies to (1), for all forms of the potential function $U(x)$ here considered. The integrated density of states is defined as the number of eigenvalues between the lowest eigenvalue and E. Therefore, if $E_m < E < E_{m+1}$, then the density of states integrated up to E is precisely equal to the integer m, i.e. the integrated state density could be determined by *counting the zeros of* $\psi(x,E)$ *within the interval* $(0,L)$.

   This zero–counting method provides a powerful tool to determine to *average integrated density of states* $N(E)$ *for random potentials* [2,8,9,10]. We apply the method to a new important class of random potentials. In addition, we develop an analogy with a rotator model of chaotic systems.

   With the abbreviations

$$\frac{2m}{\hbar^2} U(x) = V(x), \quad \sqrt{2mE}/\hbar = k, \tag{2}$$

we may introduce a new function $\theta(x)$, called the *phase*, defined by

$$\tan\theta(x) = \frac{\psi'(x)}{k\psi(x)}. \tag{3}$$

Taking the derivative of $\tan\theta$ with respect to x, we find, using (1–3) the first–order nonlinear differential equation

$$\theta' \equiv \frac{d\theta}{dx} = -k + k^{-1}V(x)\cos^2\theta. \tag{4}$$

Eq. (4) is equivalent to the Schrödinger equation (1). The phase function $\theta(x)$ has the following properties:
i)   In regions of zero potential it is

$$\theta(x) = -kx + const., \tag{5}$$

hence $\theta(x)$ is identical with the quantity commonly denoted as the phase of a real solution, e.g. that of $\sin kx$ of (1).
ii)   Due to (3), whenever the wave function ψ passes through zero, and $\psi' \neq 0$, the phase is an odd multiple of $\pi/2$:

$$\theta = \frac{\pi}{2} + n\pi, \qquad n = \text{integer.} \tag{6}$$

iii)   If the potential is a nonzero constant, $V(x) = V_0$, the differential equation
(4) can be easily solved: the solution is given by (22).
We will assume now, that $V(x)$ is a random potential. Different realizations of the
potential will lead to different functions $\theta(x)$ and different integrated state
densities.

Suppose that we observe N members of length L each of the ensemble of
random potentials. An electron be present in each potential, and we convene that
its wave function should have phase $\theta = 0$ at $x = 0$. We define a density $P(x,\theta)$ so
that the number $dN(x,\theta)$ of wave functions having phase between θ and $\theta + d\theta$ at
the point x be

$$dN(x,\theta) = N \, P(x,\theta)d\theta. \tag{7}$$

Evidently, for all x we must have

$$\int_0^\pi P(x,\theta)d\theta = 1. \tag{8}$$

The density $P(x,\theta)$ may also be viewed as the *probability density* for the phase of
the wave function in one of the random potentials to have value θ at the point x.

Suppose now that we wish to determine the average number of zeros of a wave
function in the interval $(x, x + dx)$, where the average is taken over a very large
number N of random potentials. From what was said above, the answer to this
question will be the same as to the following query: what is the number $\Delta N(x,\theta)$ of
wave functions whose phase passes through $\pi/2(\mathrm{mod}\,\pi)$ within $(x, x + dx)$? This
number is equal to the number of wave functions whose phase at the point x is in
the interval $(\pi/2, \pi/2 - \Delta\theta)$, where $\Delta\theta$ is the increment of phase of these wave
functions within Δx. For infinitesimal Δx

$$\Delta\theta = \theta'\Delta x, \tag{9}$$

hence

$$\Delta N(x,\tfrac{\pi}{2}) = NP(x,\tfrac{\pi}{2})|\,\Delta\theta| = NP(x,\tfrac{\pi}{2})|\,\theta'\,|_{\theta=\pi/2}\Delta x = Nk\,P(x,\tfrac{\pi}{2})\Delta x, \tag{10}$$

because (4) shows that whatever the potential, for $\theta = \pi/2$ the derivative of the
phase is $\theta'(\pi/2) = -k$.

It has been shown [11] that for the class of random potentials $V(x)$ we are
considering, the distribution $P(x,\theta)$ becomes stationary when $x \to \infty$ (and of course
$L \to \infty$). These potentials are characterized by the Markov property.
If the distribution $P(x,\theta)$ becomes stationary for $x \to \infty$, we may speak about the
average number of zeros of the wave function per unit length. According to
Sturm's theorem, described above, this number will be equal to the average

integrated density of states per unit length of a given random potential $V(x)$[1]. It follows then from (10),

$$N(E) = \lim_{\Delta x \to 0} \lim_{x \to \infty} \frac{\Delta N(x, \pi/2)}{N \Delta x} = k\, P(\pi/2), \tag{11}$$

where $P(\pi/2) = P(\infty, \pi/2)$. According to the ergodic principle, this density of states may also be considered as the ensemble average state density per unit length of the given statistical ensemble of random potentials of finite (but sufficient) length.

For later reference, we will now define a quantity called flux $\Phi(\theta)$. Suppose, for simplicity[2] that it is possible to define a joint probability density $P(V, \theta, x)$ such that $NP(V, \theta, x) dV d\theta$ is the number of potentials in the ensemble which at point x have a value in the interval $(V, V + dV)$, and where the phase of the wave function is in the interval $(\theta, \theta + d\theta)$. Clearly the function $P(\theta, x)$ defined in (7) is given by

$$P(\theta, x) = \int_{-\infty}^{\infty} P(V, \theta, x) dV, \tag{12}$$

and if $P(\theta, x)$ becomes independent of x for $x \to \infty$, so does $P(V, \theta, x)$. We set $P(V, \theta, \infty) = P(V, \theta)$.

The flux $\Phi(\theta)$ will now be defined as

$$\Phi(\theta) = N \int P(V, \theta) \left| \theta' \right| dV = N \int P(V, \theta) \left| -k + k^{-1} V \cos^2\theta \right| dV. \tag{13}$$

$\Phi(\theta)$ is called the flux for the following reason: if we represent for a given x each wave function $\psi(x)$ belonging to a member of the ensemble of potentials by a point in a two–dimensional "phase plane" of the values of $V(x)$ and $\theta(x)$ of the wave function, then as x progresses, these points move in the (V, θ) plane. The number of points which flow across a line $\theta = const$ per unit interval of x is the flux $\Phi(\theta)$.

3. THE CLASS OF RANDOM POTENTIALS

In this section we will choose a model one–dimensional random potential, using two criteria: (1) The potential should imitate as closely as possible a smoothly varying potential produced by a chain of randomly distributed different ionic species. (2) The potential should be mathematically tractable. From the point of view of criterion (2), the best choice would be a potential for which

$$<V(x_1) V(x_2)> = c\delta(x_1 - x_2), \tag{14}$$

---

[1]From here on we will call this quantity simply "density of states".

[2]More general definitions in terms of probability measures are possible, but will not be needed here.

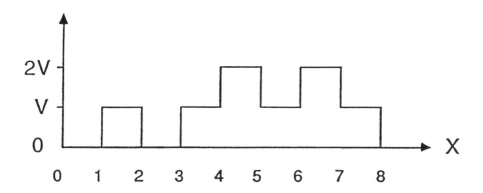

Fig.1   Illustration of a multi–step random potential. An example of a
three–step potential is shown. Proceeding from left to right, at the
points of the x–axis indicated by integers the potential may increase or
decrease by one step, or remain constant. The respective probabilities of
these changes, which are independent of x, are given by the stochastic
matrix described in the text.

because this would eliminate any correlation between values of the potential at
different points. However, such a "white–noise" potential could jump
discontinuously by any amount between arbitrarily close point of the x–axis. No
real potential exhibits such a behavior.

We chose a "random staircase" potential as follows: At the beginning of any
interval Δx, the potential will either increase one step, stay at its value, or
decrease one step with a certain probability of each of these events. For an
example, see Fig.1. In mathematical terms, set

$$x_i - x_{i\text{-}1} = \Delta x, \qquad \text{for all positive integers i; } x_0 = 0,\ x_n = L \tag{15}$$

and

$$V_m - V_{m\text{-}1} = V, \qquad \text{for } m = -M + 1,...,M;\ V_{\text{-}M} = 0. \tag{16}$$

The potential is assumed to be constant in any interval $(x_{i\text{-}2}, x_{i\text{-}1})$ and will be
denoted by $V(x_{i\text{-}1})$. Further, if $V(x_{i\text{-}1}) = V_m$ then

$$V(x_i) = \quad
\begin{array}{l}
V_{m+1} \text{ with probability } p_{m,m+1}; \\
V_m \quad \text{ with probability } p_{m,m}; \\
V_{m-1} \text{ with probability } p_{m,m\text{-}1}.
\end{array} \tag{17}$$

Since it is supposed that the potential varies between the limits $V_{-M} = 0$ and $(2M+1)V$, it has to be assumed that

$$p_{-M-1,-M} = p_{-M,-M-1} = p_{M,M+1} = p_{M+1,M} = 0. \tag{18}$$

Clearly $p_{ik} = 0$ for $i \neq k-1, k, k+1$, and

$$p_{m,m+1} + p_{m,m} + p_{m,m-1} = 1, \qquad m = -M,...,M. \tag{19}$$

The random potential so defined represents an x–homogeneous Markov chain, and if the steps V_m are represented as points on a line, the process is isomorphous to a random walk in one dimension with reflecting barriers (at V_M and V_{-M}). See, e.g. [11]. We recall, that x–homogeneity means that the transition probabilities p_{mn} are x–independent, and the Markov property signifies that the probability $r(x_i, V_m)$ of $V(x_i) = V_m$ depends only on the value of $V(x_{i-1})$, but not on the values of $V(x)$ at preceding points.

The transition probabilities p_{nm} are the elements of the so–called stochastic matrix. We only admit potential jumps to the next highest or next lowest potential. This leads to the tridiagonal stochastic matrix defined in (19). This matrix describes a correlated chain, in which atoms do not accept neighbours whose potential differ by more than one step from their own potential. One may ask whether a) the ensemble of chains defined by such a stochastic matrix will have a well–defined concentration c_i of each species of atoms and whether b) the ensemble of chains, each of which is built up from left to right adding potentials randomly through the process $V_i \rightarrow V_k$ will be identical to the ensemble built up from right to left.[3]

To calculate the concentrations c_i from a given stochastic matrix it suffices to remark, that the probability c_i that any given site is occupied by an atom of species i is equal to the sum of probabilities c_k that the previous site is occupied by an atom of species k, multiplied by the probability p_{ki} that an atom of species i will follow an atom of species k, that is

$$c_i = c_{i-1}p_{i-1,i} + c_i p_{ii} + c_{i+1}p_{i+1,i}, \quad i = -M+1,...M-1, \tag{20A}$$

and for the highest and lowest potentials

$$c_{-M} = c_{-M} p_{-M,-M} + c_{-M+1}p_{-M+1,-M}, \quad c_M = c_M p_{M,M} + c_{M-1}p_{M-1,M}. \tag{20B}$$

The solution of (20A,B) is the recursion relation

$$c_{i+1} = \frac{p_{i,i+1}}{p_{i+1,i}} c_i, \qquad i = -M,..., M-1, \tag{21}$$

where c_{-M} is determined by $\sum_i c_i = 1$ ($i = -M,...,M$). Hence the chains have well defined concentrations of species. Furthermore, the recursion relation (21) ensures automatically that the probability of occurrence of an adjacent pair $(i, i+1)$ of

---

[3]The authors are indebted to C. Itzykson for raising this latter question.

atoms of type $i + 1$ and i is the same as of an adjacent pair $(i + 1, i)$ since these respective probabilities are $c_{i+1}p_{i+1,i}$ and $c_ip_{i,i+1}$. Hence the ensemble of chains is invariant under x–reflection. This answers questions a) and b) raised above.

We now return to the probability density $P(x, \theta)$ defined in (7) and consider this function at the discrete point x_i introduced in (15). The probability of finding the phase in the interval $(\theta, \theta + \Delta\theta)$ at the point x_i depends on the whole course of the potential between x_0 and x_i. Hence $P(x_i)$ is a functional of the potential $V(x)$. However, all values of $V(x)$ for $x < x_i$ may be eliminated from $P(x_i, \theta)$, since these values of $V(x)$ serve only to determine $\theta(x_{i-1})$. Once $\theta(x_{i-1})$ and $V(x_i)$ are known, $\theta(x_i)$ may be calculated by integrating (17) between x_{i-1} and x_i to yield

$$\tan\theta(x_i) = \frac{-\alpha^{-1}\tan(k'\Delta x) + \tan[\theta(x_{i-1})]}{1 + \alpha\tan(k'\Delta x)\tan[\theta(x_i)]}, \quad k'^2 = k^2 - V(x_i), \alpha = k/k'. \tag{22}$$

This equation shows that $\theta(x_i)$ depends only on x_{i-1} and $V(x_i)$. Therefore $\theta(x_i)$ is also a Markov process. On the other hand, $\theta(x_i)$ also depends on the value of the random variable $V(x_i)$ hence $\theta(x_i)$ and $V(x_i)$ are the components of a *two–dimensional Markov process*.

On the basis of what was said above, we now introduce the *conditional probability* $P(x_i, \theta, V)dV$, *which gives the probability of finding, at the point x_i, the phase in the interval $(\theta, \theta + d\theta)$, provided that at the same point the potential has the value V.* The probability $P(x_i, \theta)$ defined in (7) is obtained from this conditional probability through

$$P(x_i, \theta) = \sum_{m=-M}^{M} P(x_i, \theta, V_m)r(x_i, V_m), \tag{23}$$

where $r(x_i, V_m)$ is the probability that $V = V_m$ at the point x_i, which is given, by virtue of (15–18) by

$$r(x_i, V_m) = \sum_{n=-M}^{M} r(x_{i-1}, V_n)p_{nm}. \tag{24}$$

Since (24) completely specifies the function $r(x_i, V_m)$ if its values $r(0, V_m)$ are given, our problem is reduced to the determination of the conditional probability density $P(x_i, \theta, V_m)$.

One may think of the combined process as a random walk in two dimensions, (θ, V), in which the probability of the θ–component of the next step depends on both the θ– and V–coordinates of the present position, whereas the V–component of the next step depends only on the V–component of the present position. Our ultimate goal is the calculation of the state density, for which the knowledge of $P(\infty, \pi/2)$ is needed. Eqs. (23) and (24) show that this goal will be achieved if we are able to calculate the conditional probability density $P(x_i, \theta, V_m)$.

Through the work of Chapman, Kolmogorov and Smoluchowski we know how to formulate a differential–difference equation for $P(x_i, \theta, V_m)$ (see, e.g. [12]). It is at this point, that the Markov–property of the potential enters, since it is not clear how to formulate such an equation for general non–Markov–type potentials. The Markov property is also used, when proving the ergodic nature of the proces, i.e.

when demonstrating, that $\lim P(x,\theta)$ for $x \to \infty$ exists and is independent of $V(x)$.

4. DIFFERENTIAL–DIFFERENCE EQUATIONS FOR THE CONDITIONAL PROBABILITY

We now formulate equations to describe the evolution of the conditional probability $P(x,\theta,V_m)d\theta$ as a function of the variable x. It is important to keep the differential $d\theta$ as a factor of P, because $d\theta$ is also a function of x. If, at the point x_{i-1}, the phase was in the infinitesimal interval $(\theta_{i-1},\theta_{i-1}+d\theta_{i-1})$, then at the point x_i it will be within the interval $(\theta_i,\theta_i + d\theta_i)$. With some algebra, it can be shown, that

$$d\theta_{i-1}^{(m)} = R(\theta,\theta_{i-1},V_m)d\theta_i^{(m)}, \tag{25}$$

where

$$R(\theta,\theta_{i-1},V_m) = \frac{1 - \beta_m\cos^2\theta_{i-1}}{1 - \beta_m\cos^2\theta_i}, \qquad \text{and } \beta_m = V_m/k^2. \tag{26}$$

We use the superscript m in (25) to indicate that $d\theta_i$ depends on V_m.

The conditional probability that at the point x_i the potential has the value V_m and the phase is in the interval $(\theta_i,\theta_i+d\theta_i)$, is the sum, over n, of the conditional probabilities that at the point x_{i-1} the potential had the value V_n and the phase was in the interval $(\theta_i,\theta_i+d\theta_{i-1})$ multiplied by the probability p_{nm} of the change of the potential from V_n to V_m. Hence we have

$$P(x_i,\theta_i,V_m)d\theta_i^{(m)} = \sum_{n=-M}^{M} P(x_{i-1},\theta_{i-1},V_n)p_{nm}d\theta_{i-1}^{(m)}. \tag{27}$$

We introduce the notation

$$\alpha_m^2 = (1 - \beta_m)^{-1} = k^2(k^2-V_m)^{-1} \tag{28}$$

and, using (22) find the somewhat congested formula

$$\Delta\theta_i^{(m)} \equiv \theta_i - \theta_{i-1}^{(m)} = \theta_i - \tan^{-1}\{\alpha_m^{-1}\tan[\alpha_m^{-1} + \tan^{-1}(\alpha_m\tan\theta_i)]\}. \tag{29}$$

Eqs.(25) and (26) allow us to cancel the factor $d\theta_i^{(m)}$ in (27) and to obtain for the unknown function P the recursion relation

$$P(x_i,\theta_i,V_m) = \sum_{n=-M}^{M} P(x_{i-1},\theta_{i-1},V_n)p_{nm}R(\theta_i,\theta_{i-1},V_m). \tag{30}$$

If we eliminate θ_{i-1} from the set of equations (30) by means of (29), then the only angle remaining in these equations is θ_i, hence for arbitrary fixed value of θ we have a set of difference equations with respect to the variable x along the chain axis.

We are interested in stationary solutions of (30) i.e. solutions such that for any θ:

$$P(x_i, \theta, V_m) = P(x_{i-1}, \theta, V_m) \equiv P(\theta, V_m). \tag{31}$$

It follows then from (30) that these stationary solutions must obey the equations

$$P(\theta, V_m) = \sum_{n=-M}^{M} P(\theta - \Delta\theta^{(m)}, V_n) p_{nm} R(\theta, \theta - \Delta\theta^{(m)}, V_m), \tag{32}$$

with $\Delta\theta^{(m)}$ given by (29) with the subscripts i dropped. These equations are of the type well known in the theory of random processes as Kolmogorov– or Fokker–Planck–equations [11,12].

5. TRANSITION TO DIFFERENTIAL EQUATIONS

The set of difference equations (32) does not lend itself to an analytic solution for $P(\theta, V_m)$. The reason for this is, that even though the variable x is increased in equal discrete steps, the variable θ increases in unequal steps $\Delta\theta^{(m)}$. Therefore it does not seem possible to obtain a closed set of equations valid for all θ. This difficulty may be circumvented by choosing Δx infinitesimally small, so that we may set

$$\lim_{\Delta x \to 0} \frac{P(\theta - \Delta\theta^{(m)}, V_m) - P(\theta, V_m)}{\Delta\theta^{(m)}} = -\frac{\partial P(\theta, V_m)}{\partial\theta}. \tag{33}$$

One can convince oneself, that (29) guarantees the existence of such a limit, if $P_m(\theta, V_m)$ is a differentiable function of θ. Indeed, in first order of Δx, Eq. (29) reduces to

$$\Delta\theta^{(m)} = -a_m k \Delta x, \qquad a_m = 1 - \beta_m \cos^2\theta. \tag{34}$$

Therefore, $\Delta\theta^{(m)}$ is proportional to Δx in (33). We shall need one more relationship to reduce (29) to differential form: starting from (26), some algebra yields

$$R(\theta, \theta - \Delta\theta^{(m)}, V_m) \simeq \frac{da_m(\theta)}{d\theta} k \Delta x. \tag{35}$$

Using an infinitesimal interval Δx along the axis imposes restrictions on the stochastic matrix p_{nm}. Since p_{nm} is the probability that at the end of the interval Δx the potential will change from V_n to V_m, we have to agree that p_{nm} be proportional to Δx unless $n = m$, otherwise we would get a wildly jumping random potential i.e. a white noise. Therefore we require

$$p_{nm} = p_{nm}^0 \Delta x, \qquad p_{mm} = 1 - p_{mm}^0 \Delta x, \qquad \Delta x \ll k^{-1}. \tag{36}$$

Eqs. (18) and (19) impose on the constants p^0_{nm} the conditions

$$p^0_{-M-1,-M} = p^0_{-M,-M-1} = p^0_{M,M+1} = p^0_{M+1,M} = 0, \tag{37}$$

and

$$p^0_{m,m+1} - p^0_{mm} + p^0_{m,m-1} = 0, \qquad m = -M,...,M. \tag{38}$$

The only restriction on the range of p^0_{nm} is that they have to be positive. Introducing (33–38) into (32) and denoting $P(\theta, V_m) \equiv P_m(\theta)$, we obtain, in first order of Δx:

$$P_{m-1}(\theta)p^0_{m-1,m} - P_m(\theta)p^0_{m,m} + P_{m+1}(\theta)p^0_{m+1,m} +$$

$$+ k\frac{d}{d\theta}[P_m(\theta)a_m(\theta)] = 0, \qquad m = -M,...,M. \tag{39}$$

This is a coupled set of first–order differential equations for the $2M+1$ functions $P_m(\theta)$. These equations have to be supplemented by the periodicity conditions for θ, and the normalization of probabilities. The former are

$$P_m(\pi) = P_m(0) \qquad m = -M,...,M. \tag{40}$$

It can be shown, that if c_m are the concentrations of the different species producing the potential V_m, where c_m are related to the elements of the stochastic matrix through (21) then

$$\int_0^\pi P_m(\theta)d\theta = c_m, \qquad \sum_{m=-M}^{M} c_m = 1. \tag{41}$$

Adding all equations (39) and making use of (38), one finds

$$\frac{d}{d\theta} \sum_{m=-M}^{M} P_m(\theta)a_m(\theta) = 0, \tag{42}$$

hence, using (35) and (10), we find the *constant of motion*

$$\sum_{m=-M}^{M} P_m(\theta)a_m(\theta) = \sum_{m=-M}^{M} P_m\left(\frac{\pi}{2}\right) = k^{-1}N(E). \tag{43}$$

It is clear from (43) that the density of states $N(E)$ can be calculated by solving (39) and (40). The reader will recall, that the constants p_{nm} are determined by the random potential, and that the trigonometric functions $a_m(\theta)$ depend on the potential as well on the energy of the particle. The solution of the differential equations (39) is rendered difficult by the presence of the trigonometric functions $a_m(\theta)$.

6. THE JOGGERS' MARATHON

A pictorial representation may be given to the process described by the differential equations (39), for fixed electron wave number k. We imagine a circular track of arbitrary radius. The phase angle θ which occurs in these equations is represented by the polar angle θ, measured from a fixed, arbitrary radius vector. The length x along the one–dimensional potential will be represented by the time x needed by a jogger, who starts at $\theta(x) = 0$, to reach the angular position $\theta(x)$. This angle is the phase of the wave function at the point x along the axis.

The instantaneous angular velocity of the jogger is given by (4), i.e.

$$\frac{d\theta}{dx} = -k + k^{-1}V(x)\cos^2\theta(x). \tag{44}$$

(Instead of the jogger, less sporting readers may also speak of a *rotator*[4]). We assign a separate jogger to each realisation of the random potential, hence there are N joggers who begin their marathon, all at the same time $x = 0$ and at the same place $\theta = 0$. With the passage of time, however, they loose phase coherence and spread out around the circle, because they speed up and slow down differently, according to the random potential in (44) to which they are assigned. The probability $P(x,\theta)d\theta$ introduced in Sec. 2 is now the probability to find any selected jogger in the angular interval $d\theta$ at time x. It follows from the ergodic property, that this probability becomes stationary as $x \to \infty$, but the stationary probability distribution $P(\theta)d\theta$ is, in general not uniform: i.e. the density of joggers varies along the circle. Clearly, in the stationary state the flux Φ, i.e. the number of joggers which pass a given cross–section of the track per unit time, will be independent of θ, the angle at which this cross–section is situated. As one can see from (44), there are two points on the track, where $d\theta/dx = -k$, hence all joggers have the same speed irrespective of the potential. These points are at $\theta = \pi/2$ and $3\pi/2$. In these points the flux Φ is easy to calculate: it is simply the density $P(\pi/2)$ of joggers multiplied by their speed k. According to (43), this quantity is the required density of states $N(E)$ which equals the average number of wave–functions passing through zero per unit time i.e. Φ.

We now elaborate this analogy further for the step–type potential discussed in Sec. 3. For ease of presentation, we choose $M = 1/2$, i.e. a potential which takes on the values 0 and V. Fig.2 shows $2M + 1$ (in this case two) concentric tracks. Each jogger starts at $x = 0$ and runs on track $V = 0$ with speed $-k$ as long as $V(x) = 0$ and jumps to track $V = 1$ when $V(x)$ changes to V. His speed is then $v = -k + k^{-1}V\cos^2\theta(x)$. He jumps back to track 0 when the potential changes to zero, and so on. The functions $P_{-1/2}(\theta)$, $P_{1/2}(\theta)$ represent the stationary densities of joggers on the two tracks. The density of states is the sum of the fluxes on the two tracks:

$$N(E) = k[P_{-1/2}(\pi/2) + P_{1/2}(\pi/2)]. \tag{45}$$

If the particle energy exceeds the maximum value of the potential, i.e. $k^2 > V$, then $d\theta/dx < 0$ for all θ, hence all joggers keeps circling clockwise. If $k^2 < V$, the joggers move in the counterclockwise direction in the two segments of the circle, where $\cos^2\theta < k^2/V$.

---

[4]Other analogies with rotators have been proposed in [13].

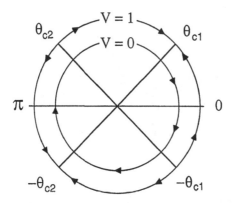

Fig.2   The joggers' marathon or rotator model illustrated in the case of a
          random potential which takes the two values $V = 0$ and $V = 1$. The
          model is described in detail in Sec.6. The angle θ_{c1} is an asymptotic
          point of approach (attractor), whereas θ_{c2} is an asymptotic point of
          withdrawal for the joggers on the track $V = 1$. The arrows indicate the
          direction of motion.

7. THE TWO–STEP MODEL

In the previous section the multi–step model has been derived and its general
aspects were presented. The present and next sections deal with the simplest cases.
In the first case the potential can take only two possible values, such as 0 and V.
This case will be called the two–step model. We will also treat the case where the
potential can take three values, 0, V and $2V$; this case will be referred to as the
three–step model.

We recall, that an n–step potential may be regarded as being produced by n
ionic species, whose relative concentrations are given, and whose atoms form a
random sequence. Each ionic species is thought to produce a constant potential
characteristic of that species, and we wish to determine the density of states of
non–interacting electrons in this potential. The equation derived in the previous
sections refer to the case where the potential can only change one step at a time.
However, equations of the same type may also be derived for the more general
case.

One may also consider the n–step model as characterizing a layered medium,
each layer having one of the n possible values of the refractive index. The layers
are stacked in a random order, but the concentration of each type of layer with a
given refractive index is given. In this case, the density of states of the
electromagnetic field (light) would be the object of study.

In the two–step model the potential can have two values, 0 and V. In the set of
equations (39) we set $M = 1/2$; the system can then be solved by quadratures. The
stochastic matrix depends on two parameters which will be denoted q_0 and q_1,
defined by

$$p^0_{-1/2,-1/2} = p^0_{-1/2,1/2} = q_0, \text{ and } p^q_{1/2,1/2} = p^q_{1/2,-1/2} = q_1. \tag{46}$$

The random potential may be thought of as being produced by two sorts of ions with concentrations

$$c_{-1/2} = \frac{q_1}{q_0 + q_1} \text{ and } c_{1/2} = \frac{q_0}{q_0 + q_1}, \tag{47}$$

respectively. The concentration $c_{-1/2}$ refers to the ions producing zero potential, whereas $c_{1/2}$ is the concentration of ions of potential V.

When k is less than $V^{-1/2}$, two angles θ_{c1} and θ_{c2} occur, for which $1-k^{-2}V\cos^2\theta = 0$. θ_{c1} and θ_{c2} will be called "critical angles". Assuming that θ is different from a critical angle and using (43), we can transform the system of equations (39) to a form which allows a certain decoupling:

$$C^{-1}P_{1/2}(\theta) = (1 - \frac{V}{k^2}\cos^2\theta)^{-1}[1 - C^{-1}P_{-1/2}(\theta)], \tag{48}$$

$$C^{-1}P_{-1/2}(\theta)\frac{d}{d\theta}f(\theta) = \frac{d}{d\theta}f(\theta) - k^{-1}q_0 + \frac{d}{d\theta}[C^{-1}P_{-1/2}(\theta)], \tag{49}$$

where

$$\frac{d}{d\theta}f(\theta) = k^{-1}[q_0 + q_1(1 - \frac{V}{k^2}\cos^2\theta)^{-1}]. \tag{50}$$

Eq. (49) is easily solvable, and the solution has the general form:

$$C^{-1}P_{-1/2}(\theta) = 1 + [C^{-1}e^{-f(\theta_0)}P_{-1/2}(\theta_0) - e^{-f(\theta_0)} +$$
$$+ k^{-1}q_0 \int_{\theta_0}^{\theta} e^{-f(\phi)}d\phi]e^{f(\theta)}. \tag{51}$$

Here, θ_0 is an arbitrary constant and C is determined by the normalization condition (41). Depending whether $V/k^2 > 1$ or $V/k^2 < 1$, the functional form of $f(\theta)$ is different:

$$f(\theta) = \begin{cases} k^{-1}[q_0\theta + q_1A\text{atan}(A\tan\theta)]; & \text{for } k^2 > V, \quad (52\text{i}) \\ k^{-1}q_0\theta + (4k)^{-1}q_1B\ln\left[(\frac{B\sin\theta - \cos\theta}{B\sin\theta + \cos\theta})^2\right]; & \text{for } k^2 > V, \quad (52\text{ii}) \end{cases}$$

where

$$A = (1 - k^{-2}V)^{-1/2}, \qquad B = (k^{-2}V - 1)^{-1/2}. \tag{53}$$

We recall that the function $P_{-1/2}(\theta)d\theta$ is the probability that the electron wave function has a phase between θ and $\theta + d\theta$ in a region of the x–axis where the potential is 0, and $P_{1/2}(\theta)d\theta$ is the probability that the phase is between θ and $\theta + d\theta$ in a region where the potential is V. Eqs. (52) indicate, that the functions $P_i(\theta)$ may behave differently, depending on whether the electron energy is higher or

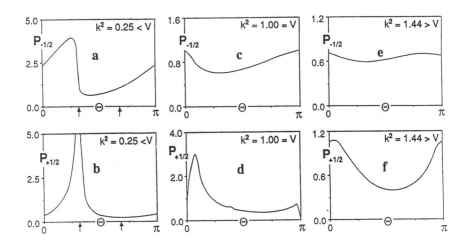

Fig.3   Plots of the probability functions $P_{-1/2}(\theta)$ and $P_{1/2}(\theta)$ for the two–step
            model for different values of the electron energy E. The height of the
            potential barriers is $V = 1$. The concentrations of the segments of
            potential 0 and V are $c_{-1/2} = 0.5$ and $c_{1/2} = 0.5$, respectively. The
            critical angles Θ_c are indicated by arrows. They depend on the electron
            wave number k. In a and b the electron energy is less, in c and d it is
            equal, and in e and f it is greater than the barrier height.

lower than the potential V. We discuss these two cases separately.

a) The electron energy is greater than the highest potential: $2m\hbar^{-2}E = k^2 > V$.
Using the general form (51) of the function $P_{-1/2}(\theta)$, the periodic boundary
condition (40) and (52i) we obtain

$$C^{-1}P_{-1/2}(\theta) = 1 + k^{-1}q_0 \left[\int_0^\theta e^{-f(\phi)}d\phi - D_1 \right] e^{f(\theta)}, \tag{54}$$

where

$$D_1 = [1 - e^{-k^{-1}(q_0+q_1A)\pi}]^{-1} \int_0^\pi e^{-f(\phi)}d\phi,$$

and $f(\theta)$ is given by (52i). The function $P_{1/2}(\theta)$ is obtained from (48) by using (54).
Fig.3 e and f presents an example of the functions $P_{-1/2}(\theta)$ and $P_{1/2}(\theta)$ for an
energy which exceeds the potential V, for a particular set of values q_0 and q_1.

b) The electron energy is less than the highest potential: $2m\hbar^{-2}E = k^2 \le V$.
In this case two critical angles appear (see also Fig.2). These angles divide the
interval $<0;\pi>$ of the variable θ into three subintervals $<0;\theta_{c1}>$, $(\theta_{c1};\theta_{c2}>$ and
$(\theta_{c2};\pi>$. For these subintervals the functions $P_i(\theta)$ must be constructed separately.
For $\theta = \theta_{ci}$ (52) yields

$$C^{-1}P_{-1/2}(\theta_{ci}) = 1, \quad i = 1,2. \tag{55}$$

Using (41), (51) and (52) we obtain the following expressions for the function $P_{-1/2}(\theta)$

$$C^{-1}P_{-1/2}(\theta) = \begin{cases} 1 + k^{-1}q_0\left[\int_0^\theta e^{-f(\phi)}d\phi + D_2\right]e^{f(\theta)}, & 0 \le \theta \le \theta_{c1}; & (56i) \\[2ex] 1 - k^{-1}q_0\left[\int_\theta^{\theta_{c2}} e^{-f(\phi)}d\phi\right]e^{f(\theta)}, & \theta_{c1} \le \theta \le \theta_{c2}>; & (56ii) \\[2ex] 1 + k^{-1}q_0\left[\int_{\theta_{c2}}^\theta e^{-f(\phi)}d\phi\right]e^{f(\theta)}, & \theta_{c2} \le \theta \le \pi; & (56iii) \end{cases}$$

where

$$D_2 = e^{q_0k^{-1}\pi}\int_{\theta_{c2}}^\pi e^{-f(\phi)}d\phi,$$

and $f(\theta)$ is given by (52ii). When k^2 increases towards V, the critical angles θ_{c1} and θ_{c2} tend towards 0 and π, respectively. When $k^2 = V$ the function $P_{-1/2}(\theta)$ is described by (56ii) for whole region $<0;\pi>$ of the argument θ. The function $f(\theta)$ then tends to the simple form $f(\theta) = k^{-1}(q_0 - \cot an\theta)$.

The functions $P_i(\theta)$ $(i = -1/2, 1/2)$ for $k^2 < V$, $k^2 = V$, and $k^2 > V$ are presented in Fig.3 for a certain choice of the parameters q_0, q_1. As shown in Sec.5, the average electronic density of states is calculated from

$$N(k) = k\left[\pi + k^{-2}V\int_0^\pi C^{-1}P_{1/2}(\theta)\cos^2\theta d\theta\right]^{-1}. \tag{57}$$

The results are presented in Fig.4 for different concentrations of the two "ionic species", i.e. of segments with potential 0 and V. We consider two limiting cases, $k^2 >> V$ and $k^2 << V$:

a) When $k^2 >> V$, a simple power series expansion in Vk^{-2} yields

$$N_{appx} = \pi^{-1}k\left[1 - \frac{1}{2}\frac{V}{k^2}c_{1/2}\right] = \pi^{-1}\sqrt{2m\,E/\hbar^2}\left[1 - \frac{1}{4}\frac{V\hbar^2}{mE}c_{1/2}\right], \tag{58}$$

where $c_{1/2}$ is the concentration of ions producing the potential V. For $k > 4.5$ the error in (58) is smaller than 1%.

b) When $k^2 << V$, the approximate formula for the average density of states is:

$$N_{appx} = \frac{q_0q_1}{q_0 + q_1}\exp(-q_0k^{-1}\pi) = \frac{q_0q_1}{q_0 + q_1}\exp\left[-\frac{\pi\hbar q_0}{\sqrt{2mE}}\right]. \tag{59}$$

This formula has been obtained in Ref.14 using a simple physical argument (see also our discussion in Sec. 9).

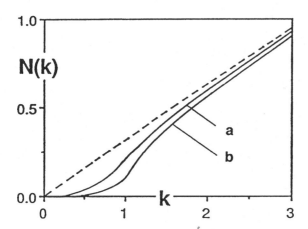

Fig.4   Average density $N(k)$ of the electronic states in the two–step model as a
           function of the free–electron wave number $k = \hbar^{-1}(2mE)^{1/2}$. The density
           of the randomly distributed segments with potential $V = 1$ is $c_{1/2} = 0.5$,
           and $c_{1/2} = 0.9$, for the curves a and b, respectively, The dotted line
           represents the free–electron case, i.e. $c_{1/2} = 0$.

8. THE THREE–STEP MODEL

We suppose now that the one–dimensional potential $V(x)$ may take on three
values, which are 0, V and $2V$. Since we are trying to model a potential which
varies as smoothly as possible within the limitations of a stepwise potential, we
assume that changes occur only by one step at a time, i.e. we exclude transitions
from 0 to $2V$ and vice versa.

Four parameters q_0, q_1, p_0, p_1 suffice to specify the stochastic matrix. These are
defined in terms of the transition probabilities as

$$p^0_{-1,0} = p^0_{-1,-1} = q_1, \qquad p^0_{0,-1} = q_0,$$

$$p^0_{0,1} = p_0, \qquad p^0_{1,0} = p^0_{1,1} = p_1. \qquad (60)$$

The subscripts –1,0 and 1 are associated with the values 0, V and $2V$ of the
potential, respectively. The corresponding set of coupled differential equations
have the form:

$$-q_1 P_{-1}(\theta) + q_0 P_0(\theta) + k\frac{d}{d\theta} P_{-1}(\theta) = 0,$$

$$q_1 P_{-1}(\theta) - (p_0 + q_0)P_0(\theta) + p_1 P_1(\theta) + k\frac{d}{d\theta}\left[(1-\frac{V}{k^2}\cos^2\theta)P_0\right] = 0,$$

$$p_0 P_0(\theta) - p_1 P_1(\theta) + k\frac{d}{d\theta}\left[(1-\frac{2V}{k^2}\cos^2\theta)P_1(\theta)\right] = 0. \qquad (61)$$

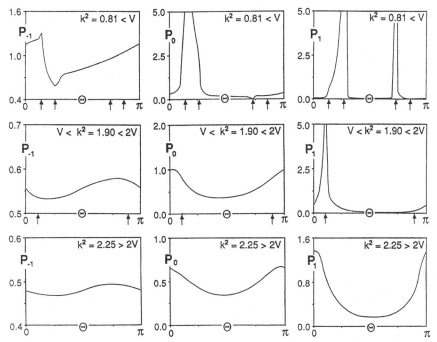

Fig. 5   Probability functions $P_i(\theta)$ $(i = -1,0,1)$ for the three–step model. In the top diagrams the electron energy k^2 is less than the lower potential barrier height V. In the middle diagrams, k^2 is between the lower and higher, $2V$, potential barriers. In the bottom diagrams, k^2 exceeds the height of the barriers. The concentrations of the segments, where the potential is 0, $V = 1$ and $2V$ are $c_{-1} = 1/3, c_0 = 1/3$, and $c_1 = 1/3$, respectively. The critical angles are indicated by arrows.

An analytical solution of the system (61) cannot be easily obtained and therefore the equations were solved numerically. Since the functions $P_i(\theta)(i=-1,0,1)$ are periodic with period π we have a two–point boundary value problem.

To find a unique solution (up to a common normalization factor) we need as many boundary conditions as there are unknown functions. Eqs.(40) provide these boundary conditions, but only when the energy of the electron exceeds the highest potential. In this case all the factors $f_n = 1 - nVk^{-2}\cos^2\theta$ $(n = 0,1,2)$ are positive for all θ, and the coefficients in the system of differential equations are non–zero. However, when the electron energy is less than the highest potential, some of the factors f_n vanish at the *critical angles* θ_c. At these angles some of the solutions P_n become infinite. The interval $(0,\pi)$ is split into subintervals by the critical angles, and the boundary conditions at $\theta = 0$ and $\theta = \pi$ do not suffice to determine a unique discontinuous solution in the subintervals between successive critical angles. A special iterative method had to be used to overcome this problem, which will be described elsewhere. The results for the functions $P_i(\theta)$, $(i = -1,0,1)$ for a set of parameters c_i, $(i = -1,0,1)$ and three values of k are shown in Fig. 5. The density of states for the same set of parameters is plotted in Fig.6.

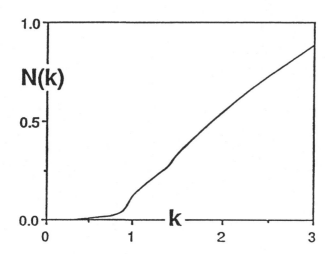

Fig.6   Density $N(k)$ of the electron states for the three–step model as a
             function of the free–electron wave number k. The concentrations of the
             randomly distributed segments with potential $0, V = 1$ and $2V$ are $c_{-1} =$
             $1/3$, $c_0 = 1/3$ and $c_1 = 1/3$, respectively.

9. DISCUSSION OF THE RESULTS

For the two–step model, Fig.3 shows the probability density of the phase θ in the
regions of zero potential (top three diagrams) and in the regions of potential V.
The probability of each value of the potential is chosen as $1/2$. In a and b the
electron energy is smaller than V. Therefore in those regions of the x–axis where
the potential is V, the phase approaches θ_{c1}. The angle θ_{c1} is an accumulation
point for the phase flow, the angle θ_{c2} is a point of flight. At θ_{c1} the probability
density $P_{1/2}$ is infinite. (Its integral is, of course, finite). Since the length of regions
of the x–axis where the phase is close to θ_{c1} is large, when the potential changes
from V to zero the phase is still close to θ_{c1}. For this reason in Fig.3a the
probability density of θ for zero potential has a maximum close to θ_{c1}. As the
electron energy increases (Figs.3c,d,e and f) the probability densities become
smoother. In fact, for the case $k^2 >> V$ (not shown), they become uniform.
     Fig.4 shows the density of electron states $N(k)$ obtained by using the
probability functions $P_{-1/2}(\theta)$ and $P_{1/2}(\theta)$ calculated numerically for the range $0 <$
$k < 3$ of the electron wave number. The two–step potential is $V = 0$ or $V = 1$,
these two values having equal probabilities for curve a. For curve b, the
probability of $V = 0$ is 0.1, that of $V = 1$ is 0.9. The density of states is a
continuous function, with no gaps. This can be simply understood, if one thinks of
the regions of zero potential as wells in between the regions of $V = 1$: The
probability distribution $\pi_0(\ell)$ of the widths ℓ of these wells is given by

$$\pi_0(\ell) = q_0 e^{-q_0 \ell}, \tag{62}$$

q_0 being proportional to the fraction of length of the x-axis where the potential is $V = 1$. According to (62) wells of all widths occur. Since the eigenstates in the wells have energies proportional to ℓ^{-2}, eigenstates with all energies will also occur and the state density has no gaps. Evidently, wide wells are less probable. Therefore the density of states is more depressed at low energies as can be clearly seen in the Fig.4. For curve b the depression is stronger, since in this case the probability of wells is smaller than for curve a.

If one applies the two–step model to optics, one has to think of a stack of two kinds of glass, with two different refractive indices. Sheets of different thickness of the two kinds of glass are stacked upon each other. The distribution of the thicknesses of the sheets of each kind is given by an equation of the type of (62). Another, equivalent interpretation of the model is to imagine that all sheets have the same –but very small– thickness, stacked upon each other according to the probability law expressed by (62). This interpretation would only be correct, if the wave length of the electromagnetic field were larger than the thickness of the sheets.

Some results, in particular analytical formulas for a two–step model, have also been obtained by other authors [14]. However, no correct state density has been found previously. Fig.5 shows the results for a three–step potential, which takes the values $V = 0,1$ and 2 with equal probabilities. In the top three diagrams the electron energy is less than 1, in the middle diagrams the energy k^2 is between 1 and 2, and in the lowest three diagrams $k^2 > 2$. When the energy is below 1 all four critical angles appear, and in the region of the x-axis where the potential is 2 the probability density $P_1(\theta)$ is infinite at two of the critical angles, whereas $P_0(\theta)$ is infinite at one of the critical angles. Between the critical angles P_1 is practically zero. Therefore $P_1(\pi/2)$ contributes very little to the density of states. When $1 < k^2 < 2$, the only discontinuity occurs at one of the critical angles in $P_1(\theta)$. $P_0(\theta)$ and $P_{-1}(\theta)$ are continuous. When $k^2 > 2$, none of the functions $P_i(\theta)$ has a discontinuity, as already discussed in Sec.8. The density of states is shown in Fig.6. As $k^2 >> V$, it approaches the free–electron density of states, but for small k it is depressed. $N(k)$ shows two pronounced bumps, which correspond to the two nonzero steps of the potential. When the concentrations c_0 and c_1 of regions with $V = 1$ and $V = 2$, respectively, are varied, the relative height of these bumps changes. It is clear, that if c_1 approaches 1 the density of states is reduced to nearly zero below $k = \sqrt{2}$.

In conclusion, we remark that the transmission properties of the random multistep potential represent another problem, which can be dealt with by the method here described. To obtain the transmittivity of such a potential situated on a finite segment of the x-axis one has to average a rather complicated function of the phase: this topic will not be treated here.

ACKNOWLEDGMENT

The support of the Swiss National Science Foundation through grant: 2000–5.295 is greatfully acknowledged.

REFERENCES

[1] R. Landauer and J.C. Helland, J. Chem. Phys. 22 (1954) 1656.
[2] H.L. Firsch and S.P. Lloyd, Phys. Rev. 120 (1960) 1175.
[3] K. Ishii, Suppl. Prog. Theor. Phys. 53 (1973) 77.
[4] P. Erdös and R.C. Herndon, Adv. Phys. 31 (1982) 65.
[5] P. Erdös, I.B.M. J. Res. Dev. 32 (1988) 47.
[6] M. Benderski' and L.A. Pastur, Sov. Phys. JETP 30 (1970) 158.
[7] M. Bôcher, *Leçons sur les méthodes de Sturm* (Paris, 1917).
[8] L.P. Gorkov, O.N. Dorokhov and F.V. Prigava, Sov. Phys. JETP 57 (1983) 838.
[9] C.J. Lambert, P.D. Beale and M.F. Thorpe, Phys. Rev. B27 (1983) 5860.
[10] A.D. Stone, D.C. Allan and J.D. Joannopoulos, Phys. Rev. B27 (1983) 836.
[11] See, e.g. W. Feller, *An Introduction to Probability Theory and its Applications*, 3rd ed. (Wiley and Sons, Inc., New York, 1968).
[12] N.G. van Kampen, *Stochastic Processes in Physics and Chemistry* (North–Holland, Amsterdam, 1981).
[13] B.U. Felderhof, J. Stat. Phys. 43 (1986) 267.
[14] I.M. Luttinger and H.K. Sy, Phys. Rev. A7 (1973) 701.

P. Erdös and *Z. Domanski* are with the Institute of Theoretical Physics, University of Lausanne, CH–1015 Lausanne, Switzerland; *Z. Domanski* is also with the Institute for Low Temperature and Structure Research of the Polish Academy of Science, Wroclaw, Poland.

CLASSICAL LIGHT WAVES AND SPINORS

Rajendra Bhandari

The analogy between the classical dynamics of a system of N harmonic oscillators and the quantum mechanics of a system with N orthogonal quantum states is discussed in the context of geometric phases. The analogy between the quantum mechanical description of a spin 1/2 system and the description of the polarization behavior of light by Jones calculus, along with the geometric descriptions known as the Bloch sphere and the Poincaré sphere are described. Finally, the elements of a more general calculus capable of handling mixed evolution of light beams in the propagation direction and polarization by means of 3×3 matrices and its usefulness in the interpretation of geometric phase experiments in optics are described.

1. INTRODUCTORY BACKGROUND

The relation between the classical and the quantum descriptions of physical systems and the transition from one to the other has intrigued physicists ever since the advent of quantum mechanics. A key feature of this problem that has emerged in recent years is the fact that the limit $\hbar \rightarrow 0$ is a necessary but not a sufficient condition for quantum phenomena to yield classical phenomena. The transition from the quantum to the classical domain is made not just by taking the limit $\hbar \rightarrow 0$, but, in addition, by a selection of a special set of basis states which in the above limit go over to the classical states. Thus quantum mechanics only allows for the possibility of classical behavior in the limit $\hbar \rightarrow 0$, but does not guarantee it. To take a simple example, a momentum eigenstate of a macroscopic object is allowed by all principles of quantum mechanics, but can never be produced in nature. Similarly, a state of the electromagnetic field with a fixed number n of photons (a Fock state) can never be produced if n is very large. This additional principle of the selection of a special basis has received considerable attention in recent literature. I wish to mention in particular the work of Zurek [1] in which this special basis is called the "pointer basis" and it is suggested that the mechanism of selection lies in the nature of the interaction between the system and its environment. More recently it was suggested [2] that for systems like the harmonic oscillator and a large spin, Zurek's pointer basis should be identified with the coherent–state basis for these systems. Coherent states are a special class of quantum states which have been studied extensively [3] and their role in the quantum to classical transition has been discussed by several authors, for example Klauder [4], Jaffe [5] and, in the context of light, by Mukunda and Sudarshan [6].

Another recent context in which the relation between the classical and the quantum descriptions of light has come up for discussion is the subject of geometric phases; in particular the manifestations of these phases in optics [7]. Very roughly speaking, the subject of geometric phases has to do with "anholonomies", i.e., situations where a system is taken through a cycle of changes

W. van Haeringen and D. Lenstra (eds.), Analogies in Optics and Micro Electronics, 69–81.

such that most of its variables return to their original values, but one of them does not. This variable is typically either the phase angle of a periodic oscillation or a wave function or an angle of rotation. One of the most interesting of these geometric phases was discovered in the context of transformations of polarization of light by S. Pancharatnam [8] way back in the fiftees. Pancharatnam showed that when a beam of polarized light is taken along a closed cycle of changes in its polarization state by means of polarization analyzers, it acquires an extra phase compared to another beam which started out in phase, traversed the same optical path but were not subjected to these transformations. This phase was predicted by Pancharatnam to be equal to half the solid angle subtended by the closed circuit of geodesics traced by the polarization state of the beam on the Poincaré sphere (which is the "state space" for the beam) at the centre of the sphere. This has been demonstrated in a series of interference experiments recently [9–11].

The recent spate of activity in the subject was sparked off by a paper by M.V. Berry [12] in which he discovered a similar "phase anholonomy" in a cyclic, adiabatic evolution of a quantum system in an energy eigenstate. Of particular interest to us in this paper is a more general formulation of the geometric phase problem by Aharanov and Anandan [13] and Samuel and Bhandari [14], following a geometric interpretation of Berry's result by B. Simon [15]. Very roughly speaking, geometric phases describe some general features of evolution of systems that depend only on the structure of the state spaces of these systems and the circuits traced out by the evolving systems in these state spaces. For instance, as pointed out in [14], if a system goes through a sequence of closely separated states ψ_0, ψ_1, ..., $\psi_n = \psi_0$ in its Hilbert space, the geometric phase is the phase of the complex number $\langle \psi_n = \psi_0 | \psi_{n-1} \rangle ... \langle \psi_2 | \psi_1 \rangle \langle \psi_1 | \psi_0 \rangle$. This number, by its very construction, is independent of the phases of the individual states involved and therefore is characteristic of a circuit in what is called the "Projective Hilbert Space", a space derived from the full Hilbert space by mapping all states that differ only in an overall phase onto a single point. Stated this way, it is easy to see why geometric phases transcend the distinction between quantum and classical and that between the equations of motion followed by different systems. For very diverse systems, one gets the same geometric phase if their state spaces are similar and they trace similar paths in these spaces (As we shall see later, classical systems can also be described by Hilbert spaces). For example, the Berry phase or the Aharanov–Anandan phase for a quantum mechanical spin 1/2 particle is exactly the same as the Pancharatnam phase which arises in the context of polarization states of classical light waves, which in turn is the same as the phase that could arise in the motion of any two–dimensional classical harmonic oscillator. These classical phases are called Hannay angles, also studied recently [16]. Similarly, geometric phases arising as a result of a light beam traversing a space curve by means of an optical fibre [17] or mirrors [18] can be understood either in terms of the Berry phase of a photon, i.e., a quantum mechanical spin 1 particle, undergoing rotations in space or in terms of the Hannay angle of a classical two–dimensional harmonic oscillator, undergoing rotations in space.

In quantum mechanics, we are very familiar with the idea of approximating complex systems with simpler representations consisting of a few base states and all relevant transformations of these systems by finite matrices which describe the behavior of these systems to a sufficient accuracy within a certain range of experimental conditions. Much of molecular physics, condensed matter physics, elementary particle physics etc. would be impossible without such an approach. In this paper we shall illustrate the usefulness of such an approach in dealing with classical monochromatic light beams of small divergence of the kind emitted by lasers. The usefulness of such an approach in dealing with polarization transformations alone has long been known since the work of Jones [19]. This is discussed in sec. 3. A recent extension of this approach to a beam of light undergoing arbitrary transformations of direction and polarization with possible mirror reflections has been found useful in understanding the geometric phase experiments in optics in a unified manner [20] and this is discussed in sec. 4. In keeping with the general aims of this volume, emphasis will be on concepts rather than mathematical rigour.

2. SOME GENERAL CONSIDERATIONS

It is a commonly used approximation in physics to describe the motion of a complex system with N degrees of freedom as a collection of N harmonic oscillators; each oscillator being a "normal coordinate" of the system. In this approximation, also called the linear approximation, any general motion of the complex system is described by specifying a set of N complex amplitudes $A_i(t)$, one for each normal mode, as a function of time. The time–dependence of $A_i(t)$ is given by, $A_i(t) = A_i(0) \exp(-i\omega_i t)$, where ω_i is the frequency of the ith normal–mode harmonic oscillator of the system. The set of N–component complex vectors: $\psi_c = $ col.$\{A_1(t), A_2(t),..., A_N(t)\}$ for all possible values of the complex numbers $A_1(t),...,$ $A_N(t)$ span an N–dimensional Hilbert space $\mathcal{H}_c$; the subscript c standing for "classical". Let us note that there is nothing "quantum" about this Hilbert space. The square of the length of the vector ψ_c, defined by $\psi_c{}^\dagger \psi_c$ represents the total energy of the system. Since a unitary transformation preserves the length $\psi_c{}^\dagger \psi_c$ all energy–conserving transformations on the system are, therefore, represented by the group of $N \times N$ unitary matrices $U[N]$ acting on the vectors ψ_c. The time dependence of this vector is given by $\psi_c(t) = \exp(-iHt)\psi_c(0)$, where H is the diagonal matrix with $H_{ii} = \omega_i$. This is nothing but a Schrödinger time–evolution for a classical system.

The electromagnetic field confined in a finite volume can also be described as a collection of $2N$ harmonic oscillators [21], where N is the number of distinct propagation modes, each with two independent states of polarization. The state of a classical electromagnetic field in vacuum can therefore be described as a vector in a Hilbert space $\mathcal{H}_c$ of $2N$ dimensions. Each component of the vector represents the complex amplitude of the harmonic oscillator corresponding to the electric field in one of the modes. The magnetic field, not being an independent quantity will be left out of the discussion. For light waves with a fixed k, $\mathcal{H}_c$ is two–dimensional. A set of two complex numbers representing the two polarization modes then completely describes the state of the field at any given instant.

A quantum mechanical system which is completely described in terms of $2N$ mutually orthogonal eigenstates of the Hamiltonian (for example a spin with total spin quantum number M such that $2M+1=2N$) is again described in quantum mechanics by a vector ψ_s in a $2N$–dimensional Hilbert space $\mathcal{H}_s$ whose length $\psi_s{}^\dagger \psi_s$, being interpreted as a total probability, is usually taken to be 1. All probability–conserving (length–conserving) transformations on the vectors ψ_s are again given by the set of all $2N \times 2N$ unitary matrices that span the group $U[2N]$. In particular, the time–evolution of the state is given by the unitary matrix $U = \exp\{(-i/\hbar)Ht\}$, where H is a hermitian matrix representing the Hamiltonian of the system. In the energy representation, H is given by the diagonal matrix $H_{ii} = E_i$, E_i being the ith energy eigenvalue of the system. The sign of the exponent of the time evolution operator U in the classical and the quantum case is purely a matter of convention. There is thus a complete analogy between a system of $2N$ classical harmonic oscillators and a quantum mechanical spin with the total spin quantum number M, such that $2M+1=2N$. The Hilbert space of all one–photon states of the electromagnetic field is another example of $\mathcal{H}_s$.

We shall next come to a slightly different question, that of the relation (not just an analogy) between the full–blown quantum Hilbert space of the electromagnetic fied $\mathcal{H}_q$ and the classical Hilbert space $\mathcal{H}_c$ which can be derived from it. In the photon number representation, also called the Fock representation, $\mathcal{H}_q$ for a $2N$ mode electromagnetic field consists of the set of mutually orthogonal vectors $|\{n_k\}> = |n_1,n_2,...,n_{2N}>$ where n_i is the number of photons in the ith mode. This is a much larger space than $\mathcal{H}_s$ discussed above. It is, however, possible to construct a $2N$ dimensional classical Hilbert space $\mathcal{H}_c$ from $\mathcal{H}_q$ via coherent states. The relation between the space of classical solutions of Maxwell's equations, the space of one–photon quantum states of the electromagnetic field and the space of coherent states of the quantum electromagnetic field has been discussed in [6]. A $2N$–mode coherent state is characterized by $2N$ complex parameters $\alpha_1,\alpha_2,...,\alpha_{2N}$ and is defined as [22]:

$$|\alpha_1,\alpha_2,...,\alpha_{2N}> = |\{\alpha_k\}> =$$

$$\sum_{\{n_k\}=0}^{\infty} \left\{ \prod_{j=1}^{2N} \exp(-\tfrac{1}{2}|\alpha_j|^2) \frac{(\alpha_j)^{n_j}}{(n_j!)^{1/2}} |\{n_k\}> \right\}. \tag{1}$$

The state $\{\alpha_k\}$, up to a normalization factor, is generated from the ground state of the $2N$ modes by the exponential operator $\exp \cdot \{\sum_j(\alpha_j a_j{}^\dagger)\}$ with $j = 1,...,2N$ where the operator $a_j{}^\dagger$ creates a photon in the jth mode. For the multi–mode coherent field, the complex vector $\{\alpha_k\}$ represents the classical complex amplitudes of the $2N$ harmonic oscillator modes and defines the classical excitation completely. $\{\alpha_k\}$ span the classical Hilbert space $\mathcal{H}_c$ discussed earlier and we emphasize that there is no difference between the evolution of this vector in $\mathcal{H}_c$ and the evolution of a $2N$–state quantum system in $\mathcal{H}_s$ if the quantum system has energy eigenvalues given by $\hbar\omega_j$.

Classical experiments with light represent transformations within the space of the coherent states. The structure of the exponential operator generating the coherent states from the vacuum makes it obvious that a unitary transformation U made on the vector col $\cdot \{a_1{}^\dagger,..., a_{2N}{}^\dagger\}$ and a transformation $U^\dagger$ made on the vector col $\cdot \{\alpha_1{}^*,...,\alpha_{2N}{}^*\}$ represent the same transformation on the coherent state. This accounts for the fact that a geometric phase seen in classical experiments can be interpreted either as a Berry phase or Aharanov–Anandan phase of a photon due to its evolution in the one–photon Hilbert space or as a Hannay angle due to evolution in the classical Hilbert space. This can be demonstrated in yet another way for experiments in which a light beam is taken around a space curve as in [17] and [18]. Consider the normalized coherent state of a single circularly polarized mode, written as:

$$|\alpha> = \exp(-\tfrac{1}{2}|\alpha|^2) \sum_{n=0}^{\infty} (n!)^{-1/2}\alpha^n|n>, \tag{2}$$

where $|n>$ is a state with n circularly polarized photons. This state $|n>$ is equivalent to a quantum mechanical particle with spin n along the direction of propagation, hence acquires a Berry phase factor equal to $\exp\{in\Omega(C)\}$. It is easy to see from (2) that the amplitude of the state $|n>$ acquiring a phase $\exp\{in\Omega(C)\}$ is equivalent to replacing α by $\alpha' = \alpha \exp\{i\Omega(C)\}$. But as we have seen earlier, α represents the complex amplitude of the classical electromagnetic wave. It follows therefore that a photon acquiring a geometric phase equal to $\Omega(C)$ is equivalent to the classical wave acquiring a geometric phase equal to $\Omega(C)$. An argument exactly similar to the one above would show that a graviton, which is a spin 2 particle, hence acquiring a Berry phase equal to $2\Omega(C)$ in going around a circuit in space is equivalent to a classical gravitational wave acquiring a geometric phase equal to $2\Omega(C)$.

3. POLARIZATION OF LIGHT AND SPIN 1/2

We have already seen that a classical two–dimensional harmonic oscillator with eigenfrequencies ω_1 and ω_2 has the same classical mechanics as the quantum mechanics of a two–state quantum system like a spin 1/2 with energy eigenvalues $\hbar\omega_1$ and $\hbar\omega_2$. Is there any paradox in a spin 1 particle like a photon behaving as a spin 1/2 particle? There is in fact none. Take any spin with a spin quantum number M. It is described by a vector in a $2M+1$ dimensional Hilbert space:

$$\psi_s = \text{col} \cdot \{c_M, c_{M-1}, ..., c_{-M+1}, c_{-M}\}; \sum_{i=-M}^{M} |c_i|^2 = 1 . \tag{3}$$

If we could find a way to constrain the $2M-1$ coefficients c_i with $i = M-1$ to $-M+1$ so that they are always zero, the system is constrained in a two–dimensional Hilbert space and all transformations on the space can be made by 2×2 matrices instead of $(2M+1)\times(2M+1)$ matrices and the system will be isomorphic to the spin 1/2. In this sense, not only photons but spin 2 particles like gravitons; in fact all massless particles would behave as a spin 1/2 particle. The reason for the existence of this constraint for massless particles is, however, a different and involved issue that we shall not go into here. It is true, however, that if gravity waves could be subjected to polarization transformations, they would also exhibit a Pancharatnam phase equal to half the solid angle on their Poincaré sphere (Fig.1). Let us emphasize, however, that the analogy between spin 1/2 and the polarization states extends only to their two–state aspect and not to the quantum statistics followed by the two particles.

A complete description of a two–state system in a pure quantum–mechanical state is given by a two–component complex vector $\text{col} \cdot \{c_1, c_2\}$, which we shall call a spinor, where c_1 and c_2 are complex numbers representing the probability amplitudes for the system to be in the states 1 and 2 respectively, such that $|c_1|^2 + |c_2|^2 = 1$. When dealing with situations where non–unitary evolution is allowed for, one might usefully consider vectors which do not satisfy the unit–norm condition [14].

Any observable M with regard to the system is represented by a 2×2 hermitian matrix and can be written in the form

$$M = A \cdot 1 + I \cdot \sigma , \tag{4}$$

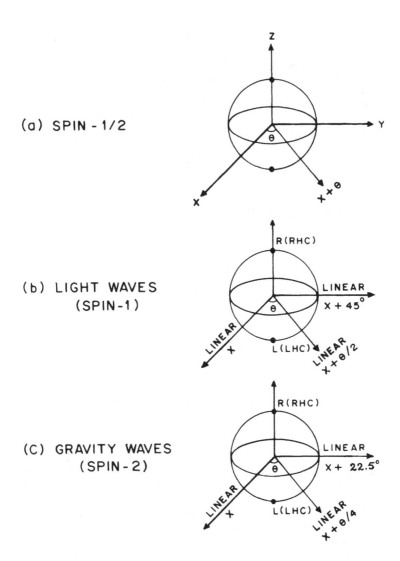

(a) SPIN - 1/2

(b) LIGHT WAVES
    (SPIN-1)

(C) GRAVITY WAVES
    (SPIN-2)

Fig.1   The projective Hilbert space for (a) a quantum mechanical two–state
             system, e.g. a spin 1/2, where X, Y, Z can be looked upon as directions in
             real space; (b) the polarization states of a photon (spin 1) or light
             waves. The two poles represent right–hand and left–hand circular
             polarizations, a point on the equator with azimuth θ represents linear
             polarization at an angle $\theta/2$ from the x–axis and all other points
             represent elliptical polarization. (c) The polarization states of a graviton
             (spin 2) or gravity waves. In this case, a rotation through an angle about
             the polar axis represents a rotation through $\theta/4$ about the direction of
             propagation in real space.

where **1** is the unit matrix, $\sigma_x, \sigma_y, \sigma_z$ are the three Pauli matrices and I_x, I_y, I_z are three real numbers so that I can be treated as a vector in real space. In particular, the density matrix for a pure, normalized state of the system can be written as the hermitian matrix $\rho = (1/2)\{1 + \mathbf{n} \cdot \boldsymbol{\sigma}\}$, where $\mathbf{n}$ is a unit vector and ρ satisfies the conditions (i) $\mathrm{tr}\rho = 1$ and (ii) $\rho^2 = \rho$. The significance of the density matrix of a pure state is that a given ρ represents all states of the system which differ only in an overall phase. The space of all density matrices form what is known as the "projective Hilbert space" of the system and is much discussed in connection with geometric phases. In the case of the two–state system, this happens to be the two–dimensional surface of a sphere in three dimensions. If one wished to consider non–hermitian operators pertaining to the system, one could allow the components of I to be complex.

Any unitary transformation on the two–state system can be represented by a 2×2 unitary matrix U, which can always be written in the form $U = \exp\{iK\}$, where K is a hermitian matrix, which in turn can be written in the form given in (4), using Pauli matrices. When the sum of the eigenvalues of K is zero, the determinant of U equals 1 and the group of such matrices forms the SU(2) group. From any U matrix that does not belong to the SU(2) group, one can construct, by suitably choosing a real number ϕ, a matrix $U' = \exp\{i\phi\} U$ such that U' belong to the SU(2) group. If we are interested in defining transformations on states only upto an overall phase, it is enough to deal only with the SU(2) matrices. For example the group of transformations that takes each density matrix defined earlier into every other density matrix is the SU(2) group. Some particularly interesting examples of unitary transformations are, (a) the time evolution operator that takes the state vector at time 0 to the vector at time t, given by $T = \exp\{-(i/\hbar)Ht\}$ and (b) the rotation operator $R(\mathbf{n}, \theta) = \exp\{-(i/2)\sigma \cdot \mathbf{n}\theta\}$, which rotates any quantum state of the system about the axis defined by the unit vector $\mathbf{n}$ through an angle θ in *real* space. It follows from the discussion above that for a spin 1/2, these two operators are the same upto an overall phase. Any time evolution of a two–state system can therefore be looked upon as the rotation in space or the "precession" of a spin about a direction in space. It is a useful analogy to think of the equivalent spin as endowed with a magnetic moment and of the Hamiltonian as a magnetic field acting on this magnetic moment. At this point, let us note a very important fact about the spin 1/2 system. Starting with a spin 1/2 in a given state, applying all possible "space rotations", by means of the 2×2 unitary matrices $R(\mathbf{n}, \theta)$ that act in the Hilbert space, one can generate all possible states of the spin 1/2 upto a phase. This cannot be done for any spin larger than 1/2. Thus for spin 1/2, spins pointing in all possible directions exhaust all physically distinct states of the system, whereas for higher spins, this is not true. For example, by rotating a linearly polarized photon (a spin 1 particle) in real space, one gets a linearly polarized photon travelling in a different direction, but one can never get a circularly polarized photon by rotations, or even by "space–reflection" operations. Stated another way, only for spin 1/2 do the rotations in real space, represented by the matrices $R(\mathbf{n}, \theta)$ exhaust all physically distinct SU transformations.

In the context of the polarization states of light, a representation exactly similar to the one for the spin 1/2 discussed above was developed by Jones [19] and is known as the Jones calculus. A plane electromagnetic wave with a fixed wave vector is represented by a vector with two complex components representing the two orthogonal components of the oscillating electric vector of the wave. Optical components like wave retarders, rotators etc. which change the polarization state

of the wave without changing the total intensity of the wave are represented by
2×2 unitary matrices while elements like a polarizer which change the total
intensity of the wave are represented by 2×2 non–unitary matrices; all of them
being called Jones matrices. The change of the polarization state of the wave due
to its passage through a sequence of such elements is computed by multiplying
several such matrices in the correct sequence.

There are many applications in quantum mechanics as well as in polarization
optics, where a geometrical representation of the states and their transformations
is much more useful. As stated earlier, every state of the two–state system, upto a
phase, can be represented by a point on the surface of a sphere in three dimensions.
For a spin 1/2 system this sphere can be looked upon as the sphere of directions in
real space. In this context, it has also been known as the Bloch sphere. In the
context of the polarization states of light, this has been known as the Poincaré
sphere. There exists a similar sphere for every massless particle, for example for
spin 2 particles like gravitons.

Although the physically distinct states of all the systems discussed above have
the same representation, the physical appearance of these states in real space are of
course different. Fig.1 shows the significance of the various points on the sphere in
case of three systems: a quantum mechanical spin 1/2 particle, light waves and
gravity waves. It is worth noting that a rotation through an angle θ about the
polar axis of the sphere is equivalent, in real space, to a rotation through an angle
θ in case of the spin 1/2 , through an angle $\theta/2$ in case of light waves and through
$\theta/4$ in case of gravity waves about the "z–axis" which means the axis of
quantization. For light waves and gravity waves this is the direction of
propagation. A very clear discussion of this point is contained in Misner, Thorne
and Wheeler [23].

Every unitary transformation (the SU(2) part) on the state of the system can
be represented by a rotation of the state sphere about an axis defined by the unit
vector n that joins the centre of the state sphere to some point on the sphere by a
certain angle, i.e. each point on the state sphere follows a circular arc. Thus, wave
retarders acting on the polarized light are the analogs of the magnetic field acting
on the spin 1/2 particle with a magnetic moment. The usefulness of a geometrical
representation in terms of the Poincaré sphere has been demonstrated in the
extensive work of Pancharatnam [8] that led to the discovery of the Pancharatnam
phase.

More recently, the power of the Poincaré sphere and of geometrical phase ideas
in general was demonstrated in another piece of work [24] in which it was shown,
using geometrical methods, that any arbitrary element of the polarization
transformation group, i.e., a rotation through an arbitrary angle along an arbitrary
direction on the Poincaré sphere can be realized by means of a single device
consisting of two quarter–wave plates and two half–wave plates in a configuration
$QHQ^{-1}H$ by merely changing the relative orientations of the principal axes of these
plates. It was also shown that a simple device consisting of a half–wave plate
sandwiched between two identically oriented quarter–wave plates (QHQ) can act
as a variable linear retarder that can be used to introduce a chosen value of phase
difference between any pair of orthogonal linear polarizations. Geometric phase
ideas played an essential role in this synthesis.

4. LIGHT WAVES AND SPIN 1

In this section we shall deal with those aspects of light waves which display spin 1 behavior. Consider the experiments [17,18] in which a beam of linearly polarized light is taken along a space curve such that it goes through a closed circuit on a sphere of directions and one observes a rotation of the plane of polarization by an angle equal to $\Omega(C)$, where $\Omega(C)$ is the solid angle of the circuit on the sphere of directions. Jordan [25] has shown that Berry's result for the geometric phase of a spin σ follows from simple properties of the rotation group. A discrete version (finite number of rotation elements) of Jordan's results would state: if you start with a vector pointing in a direction k_0, apply to it a series of rotations $R(n_1, \theta_1)$, $R(n_2, \theta_2),...,R(n_n, \theta_n)$ that take the vector to $k_1, k_2,...,k_n = k_0$ (where $R(n_i, \theta_i)$ represents a rotation about an axis normal to the plane containing k_{i-1} and k_i, θ_i is the angle between k_{i-1} and k_i), such that the "state point" on the state space, namely the sphere of directions, describes a closed circuit consisting of n *geodesic* arcs, then the product of these n rotations is equivalent to a single rotation about an axis defined by k_0, by an angle equal to the solid angle subtended by the circuit at the origin of the sphere. If k represents the axis of quantization of a spin (say the z–axis), the net result of the above product of rotations on a state $|S=N$, $S_z = \sigma>$ is a rotation about z through an angle $\Omega(C)$, which we know is equivalent to multiplying its wave function by $\exp i\{\sigma\Omega(C)\}$. What is involved here is a product of a series of $(2N+1)\times(2N+1)$ unitary matrices that represent rotations in ordinary space. The product rotation matrix $R(z, \Omega(C))$ is the diagonal matrix R_{mm} $= \exp[\{i(N-m+1)\}\Omega(C)]$, m running from 1 to $2N+1$.

   The case of the photon corresponds to $N=1$ and the Berry phase results from a product of the 3×3 spin rotation matrices acting on a three–component spinor which, in a basis in which the axis of quantization is along k, is written as $\text{col} \cdot (c_1, 0, c_{-1})$. The set of all polarization transformations on a light wave with a given k discussed in section 3 can also be written as 3×3 block–diagonal matrices S_{ij}, with $S_{22} = 1$ and S_{ij} for $i,j = 1,3$ forming a 2×2 Jones matrix. It is strongly suggestive therefore that mixed transformations of direction and polarization be described using a combination of the above two types of 3×3 matrices. In such a representation [20], the two groups of geometric phase experiments, including those involving mirror reflections can be understood in a unified manner.

   A monochromatic, nearly parallel light beam of the type emitted by a mono–mode laser can, to a good approximation, be treated as a plane wave with a wave vector k. In a general experiment, this beam is made to undergo arbitrary changes of direction and polarization by means of mirror reflections, twisted optical fibres, retardation plates and polarizers. We will ignore for the time being experiments involving active elements like optical phase conjugators as in the experiments of Ref.11. There has been a lot of effort in recent literature at constructing restricted state spaces for understanding the observed geometric phases in such experiments, particularly those involving mirror reflections [11,18,26]. These, however, either work only for ideal reflections or only for one–dimensional problems. Our representation, which overcomes these limitations is based on the general idea that the full state space of interest in these experiments is a direct product of the momentum space and the polarization space of the photons. In this representation, one seeks to represent photons by three–component spinors and transformations of the photons by a subgroup of the group of spin 1 operators. The main ingredients in this scheme are:

   (*i*) Light beams are represented by three–component spinors (spin 1 wave functions). In a basis in which the axis of quantization is along the direction of propagation these are of the form $\text{col} \cdot \{c_1, 0, c_{-1}\}$.

   (*ii*) The change of direction of a light beam from a wave vector k_{i-1} to k_i is represented by a 3×3 spin 1 rotation matrix $R(n_i, \theta_i)$ corresponding to a rotation of the spinor in real space by the angle θ_i between k_{i-1} and k_i about an axis along the normal n_i to the plane containing k_{i-1} and k_i.

(*iii*) A polarization transformation corresponding to an element placed in the initial section of the beam is represented by a block–diagonal matrix S, where $S_{22}=1$ and S_{ij} for i, $j=1,3$ is an element of the SU(2) polarization transformation group for intensity–preserving transformations and an appropriate non–unitary Jones matrix for elements like polarizers that change total intensity.

(*iv*) The polarization–transforming element S, when "parallel–transported" to the ith section of the beam is represented by the matrix

$$S' = R(n_i, \theta_i) ... R(n_1, \theta_1) S R(n_1, \theta_1)^{\dagger} ... R(n_i, \theta_i)^{\dagger} . \tag{5}$$

(*v*) The state of the beam after a sequence of rotations and polarization transformations is given by the product of the transformation matrices; an R matrix for each rotation of the beam and a suitably transformed S matrix given by (5) in the order in which they are encountered, acting on the initial spinor. When the final direction of propagation k_n is the same as the initial direction k_0, the resulting spinor represents the final state of the beam expressed in its "natural basis". This is in general different from the initial spinor even if no S elements were present in the light path, as in the experiments of Ref.17. When $k_n \neq k_0$, a final rotation matrix that changes the axis of quantization from k_0 to k_n must be applied to express the state in its natural basis.

There are two important facts that make this representation useful and practical in analyzing the results of experiments involving evolution of light beams of the above type [20]:

(1) A mirror reflection is, in general, equivalent to a product of two operations: (*i*) a rotation by the angle between the incident and the reflected beams about an axis normal to the plane containing the two beams and (*ii*) a suitably chosen wave–retarder (an SU(2) element) which depends upon the physical nature of the reflecting surface. For an ideal metal–mirror reflection, (*ii*) is a half–wave plate with one of its principal axes normal to the plane containing the incident and the reflected beams. Under these ideal conditions the action of a mirror reflection can be described as a "space reflection" operation and the restricted state spaces like the modified momentum space [26] and the space of spin directions [18] can be used. In general, however, the state space is the space generated by applying all possible spin rotation operators R to the spinor col·$\{c_1, 0, c_{-1}\}$.

(2) The final state of the beam remains unaffected if a polarization transforming element is parallel–transported to any other part of the circuit, provided the relative order of such elements remains unchanged. In particular, all such elements including those introduced in lieu of mirror reflections can be transported to the beginning of the light path so that the total evolution can be decomposed into two separate parts, the first one involving only the evolution of the beam on the Poincaré sphere and the second part involving only a parallel transport of the state resulting from the first part. Using the above "decomposition rule", for any complex configuration of optical elements one can construct an equivalent configuration which can be analyzed much more simply.

In terms of the spin 1 operators, the decomposition is justified if we note that, since R is a unitary operator,

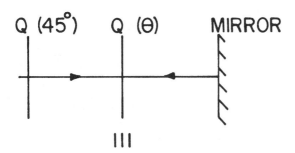

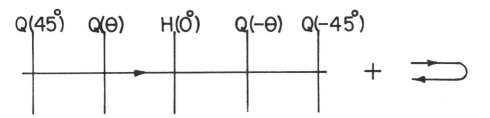

Fig.2   Decomposition of a configuration of optical elements corresponding to
the experiments of Ref.10 into an equivalent configuration. $Q(\theta)$ and
$H(\theta)$ stand for a quarter–wave plate and a half–wave plate, where θ is
the angle between the principal axis of the wave plate and a direction
normal to the plane containing the incident and the reflected beams. For
a strictly one–dimensional situation, this plane can be chosen
arbitrarily.

$$(RSR^\dagger)R|\psi> = RS|\psi> ,\tag{6}$$

where the right–hand side represents a polarization–transforming element located
in the first section of the beam, while the left–hand side represents the same
element, parallel–transported, located in the second section of the beam.
    The two special cases: (i) the phase due to an adiabatic circuit in k–space, i.e.,
an adiabatic evolution of the beam through an optically isotropic medium and (ii)
the Pancharatnam phase come out naturally when only the R or the S operators
respectively are involved. In the second case the representation reduces to the
Jones calculus.
    As an example of how the decomposition rule simplifies analysis of light
propagation problems, consider the following question: using only discrete
metal–mirror reflections, could one construct a device that would rotate an
arbitrary incident linear polarization by a fixed angle? In the above scheme, each

of the n metal–mirror reflections, assumed ideal, is replaced by a half–wave plate, which is parallel–transported to the beginning of the light path. The equivalent train therefore consists of n half–wave plates followed by a space curve which follows the actual path of the beam. The latter, as we know, rotates the electric vector by an angle equal to the solid angle on the space of directions, as demonstrated in the experiments of Ref.17. If, therefore, the collection of n half–wave plates can be made to add to "nothing", we have the required device. It is easy to see that this cannot be done with an odd number of half–wave plates (try circular polarization). On the other hand with an even number of half–wave plates, i.e. an even number of mirrors, this *can* be done. In the experiment of Chiao *et al.* [18], with four mirrors in each arm of the interferometer, such a device has in fact been achieved in each arm.

A few applications of the above scheme in describing geometric phase experiments, particularly those involving mirror reflections, have been described in [20]. In case of the Pancharatnam phase experiments involving mirrors, the use of the above scheme enables one to describe the observed non–zero geometric phases in terms of circuits on the conventional Poincaré sphere, thus obviating the need for a "Generalized Poincaré Sphere" [11]. In case of the momentum space experiment with mirrors [18], it yields a simple explanation for the cancellation of the effect of mirror reflections and enables an interpretation of the observed phase as due to a circuit in ordinary momentum space. Fig.2 shows an example of an equivalent train constructed using the decomposition rule for a one–dimensional situation like that encountered in the experiments of Ref.10.

ACKNOWLEDGEMENTS

I thank B.R. Iyer, R. Nityananda, J. Samuel and C.S. Shukre for useful discussions on the subject of this article.

REFERENCES

[1]   W.H. Zurek, Phys. Rev. D24 (1961) 1516; *ibid* D26 (1982) 1862.
[2]   R. Bhandari, Proceedings of *Meeting on the Philosophical Foundations of Quantum Mechanics*, New Delhi, March 24–26, 1988, ed. R. Nair (World Scientific), in press.
[3]   J.R. Klauder and B. Skagerstam eds., *Coherent States* (World Scientific, Singapore 1985).
[4]   J.R. Klauder, J. Math. Phys. 4 (1963) 1058.
[5]   L.G. Jaffe, Rev. Mod. Phys. 54 (1982) 407.
[6]   N. Mukunda and E.C.G. Sudershan, Pramana–J.Phys. 27 (1986) 1.
[7]   R.Y Chiao, Talk given at the 6th Rochester Conference on Coherence and Quantum Optics, June 26, 1989 and references therein.
[8]   * S. Pancharatnam, Proc. Indian. Acad. Sci. A44 (1956) 247; Collected works of S. Pancharatnam (Oxford 1975).
[9]   R. Bhandari and J. Samuel, Phys. Rev. Lett. 60 (1988) 1211; R. Bhandari, Phys. Lett. A133 (1988)1.
[10] R. Simon, H.J. Kimble and E.C.G. Sudarshan, Phys. Rev. Lett. 61 (1988) 19; T.H. Ckhyba, L.J. Wang, L. Mandel and R. Simon, Opt. Lett. 13 (1988) 562.
[11] W.R. Tompkin, M.S. Malcuit, R.W. Boyd and R.Y. Chiao, *Time Reversal of Berry's Topological Phase by Optical Phase Conjugation*, preprint.

[12]   * M.V. Berry, Proc. Roy. Soc. London A392 (1984) 45.
[13]   * Y. Aharanov and J. Anandan, Phys. Rev. Lett. 58 (1987) 1593.
[14]   * J. Samuel and R. Bhandari, Phys. Rev. Lett. 60 (1988) 2339.
[15]   * B. Simon, Phys. Rev. Lett. 51 (1983) 2167.
[16]   * J.H. Hannay, J. Phys. A18 (1985) 221.
[17]   R.Y. Chiao and Y.S. Wu, Phys. Rev. Lett. 57 (1986) 933; * A. Tomita and
       R.Y. Chiao, Phys. Rev. Lett. 57 (1986) 937.
[18]   R.Y. Chiao, A. Antaramian, K.M. Ganga, H.Jiao, S.R. Wilkinson and H.
       Nathel, Phys. Rev. Lett. 60 (1988) 1214.
[19]   R.C. Jones, J. Opt. Soc. Am. 31 (1941) 488; *Polarized Light*, ed. W. Swindell
       (Dowden, Hutchinson and Ross, 1975) contains a collection of several papers
       by Jones on the subject.
[20]   R. Bhandari, Phys. Lett. A 135 (1989) 240.
[21]   R. Loudon, *The Quantum Theory of Light* (Oxford, 1973), p.126.
[22]   J.R. Klauder and E.C.G. Sudershan, *Quantum Optics* (Benjamin, 1968),
       p.132.
[23]   C.W. Misner, K.S. Thorne and J.A. Wheeler, *Gravitation* (W.H. Freeman,
       1973), p.952.
[24]   R. Bhandari, Phys. Lett. A138 (1989) 469.
[25]   T.F. Jordan, J. Math. Phys. 29 (1988) 2042.
[26]   M. Kitano, T. Yabuzaki and T. Ogawa, Phys. Rev. Lett. 58 (1987) 523.

* Reproduced in *Geometric Phases in Physics*, eds. A. Shapere and F. Wilczek
(World Scientific, 1989).

Rajendra Bhandari is with the Raman Research Institute, Bangalore – 580 080,
India.

PART II

COHERENT OPTICS

TOWARDS OBSERVATION OF ANDERSON LOCALIZATION OF LIGHT

M.P. van Albada, A. Lagendijk and M.B. van der Mark

Localization of waves is associated with the presence of interference effects in multiple scattering. In this paper we discuss light scattering experiments on strongly scattering random media consisting of small dielectric particles (TiO$_2$, polystyrene) in a medium (air, water). Experimental observations include observation of interference in backscattering (weak localization) and study of transport. Experiments are compared with theories, and effects which are the consequence of the vector nature of light are explained. On the basis of experimental results we speculate about the possibility to observe strong localization of light.

1. INTRODUCTION

1.1 Localization

Lozalization of light is a fascinating new field in which modern developments of condensed matter physics are applied to the well–established field of linear optics. Localization as introduced by Anderson [1] refers to a dramatic change in the propagation of an electron when it is subject to a spatially random potential [2].

Several approaches can be used to describe the phenomenon of localization in a disordered medium. One method is to look at the transport properties of the electron. In this picture localization is concerned with the vanishing of the diffusion coefficient. This view of localization was given a thorough foundation by Götze [3], and later workers [4], who used transport equations of the mode–coupling type. At localization the interference of scattered waves cannot be neglected any more, but indeed its influence becomes essential ("there is life after a mean free path").

The dimension of the disordered medium is a crucial parameter. In 1–D and 2–D any degree of disorder will lead to a finite localization length while in 3–D a certain critical degree of disorder is needed before localization will set in. In the latter case the heuristic Ioffe–Regel criterion $\lambda_{mf} \lesssim \lambda$, modified by Mott [5] to

$$\lambda_{mf} \lesssim \lambda/(2\pi), \tag{1}$$

(in which λ_{mf} is the mean free path of the electron, and λ is its wavelength) applies.

Although the majority of theories of localization has been developed for the Schrödinger wave equation, it is quite clear that the concept is much broader. In principle for almost any wave equation localized solutions can be obtained when

85

W. van Haeringen and D. Lenstra (eds.), Analogies in Optics and Micro Electronics, 85–103.

solved for a random medium. This becomes particularly clear when one realizes that the Ioffe–Regel criterion can be applied to all wave phenomena. Localization of electromagnetic waves is of particular interest because one deals with localization of *vector* waves, described by a well–established and well–studied set of equations basically different from the Schrödinger wave equation.

1.2 Electron Scattering and Light Scattering

At the basis of localization is elastic scattering. In electron scattering the long–wavelength limit is s–wave scattering which is spatially isotropic and wavelength independent. The long–wavelength (Rayleigh) limit for light scattering is p–wave scattering which, although there is symmetry with respect to forward and backward scattering, is not isotropic. Moreover, the cross–section for Rayleigh–scattering varies with the inverse–fourth power of the wavelength. The very important long–wavelength limit in localization is therefore totally different for electrons and light.

The dielectric media that scatter light most strongly are those containing scatterers whose size is comparable to the wavelength. In these conditions, scattering is highly anisotropic, and forward scattering predominates. As a result, it takes several scattering events to attain a complete loss of memory of original direction. The crucial parameter here no longer is the scattering mean–free path λ_{sc}, but the transport mean–free path λ_{tr}, which may be thought of as the distance to be travelled to attain this loss of memory. λ_{sc} and λ_{tr} are related by $\lambda_{tr}= \lambda_{sc}/(1-<\cos\theta>)$, where $<\cos\theta>$ is the average cosine of the angle of scattering.

Inelastic scattering causes loss of coherence so that interference and hence localization effects are affected. In the case of electrons, inelastic scattering amounts to changing the energy of the electrons, while their number is conserved. The most important inelastic process in light propagation, however, is absorption, and amounts to removing intensity from the beam.

Electrons have a mutual interaction which is an important complication. In classical linear optics, light–light interaction is absent and a single–wave theory is exact.

Electrons are described by a scalar wave function or, if spin–dependent interactions are involved, by a two–component spinor. Light is described by a three–dimensional vector field.

Experimentally, electron localization is studied through the electrical conductance of a sample. An important aspect is the possibility to influence the interference by applying a magnetic field. In light scattering the net flux is not easily determined and, under accessible experimental conditions, the interference can not be influenced. On the other hand, it is possible to perform angular–resolved and time–resolved scattering experiments with light.

2. WEAK LOCALIZATION

2.1 Interference and Weak Localization

Anderson localization is the vanishing of the diffusion constant as a result of interference once the mean free path becomes shorter than some critical value. Now if the mean free path is not yet short enough to make the diffusion constant vanish, one may already observe interference effects that affect this constant. Let us look at interference effects in multiple scattering: A well known phenomenon in

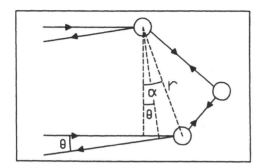

Fig.1   Time–reversed pair of light paths. In case of exact backscattering the
paths are equally long. In all other directions there will be a phase
difference between the time reversed waves.

the scattering of coherent light by a rigid random medium is "laser speckle", which
results from interference between scattered waves that traveled through the sample
along different light paths (by "light path" we mean a sequence of scattering
events $s_1, s_2, ..., s_n$). If the scatterers are allowed to move over distances of the order
of the wavelength of the light or more at the time scale of the measurement, the
spatial distribution of the scattered intensity will be the average of a rapidly
changing speckle pattern, and therefore essentially flat. Liquid random samples in
which the scatterers are subject to Brownian motion, or rigid random samples that
are spun, yield scattered–intensity patterns that almost look as they would if
interference between light paths did not exist. One type of interference, however,
survives: In the direction of pure backscattering, the waves that travel along the
same light path in opposite directions (time–reversed paths) will always have the
same phase and interfere constructively (actually this holds only for the
backscattered light component polarized parallel to the incident beam, but for the
moment we neglect polarization effects). Moving away from this direction, a
difference in phase $\Delta\phi$ develops, which increases with the angle θ between the
incoming and outgoing wave vectors and further depends on the relative positions
of the first and last scatterer s_1 and s_n, according to

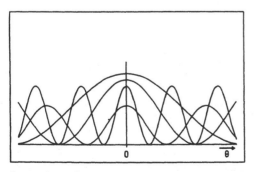

Fig.2   A number of two–point–source interference patterns belonging to
different time–reversed pairs of waves.

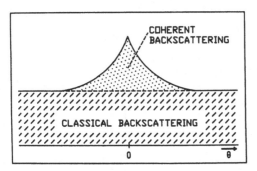

Fig.3   Sum of the two–point–source interference patterns of all contributing
             time–reversed pairs of light paths.

$$\Delta\phi = \frac{r}{\lambda}\{\sin(\alpha) - \sin(\alpha-\theta)\} = 2\frac{r}{\lambda}\cos(\alpha-\theta/2)\sin(\theta/2), \qquad (2)$$

where r is the distance between s_1 and s_n (c.f. Fig.1). With increasing θ the
interference for an individual light path will oscillate between constructive and
destructive. At $\theta=0$ all time–reversed pairs interfere constructively, at small θ only
light paths with large values of r will interfere destructively, at large θ
constructive and destructive interference will average out (c.f. Fig.2). Thus, within
a certain solid angle around the direction of pure backscattering there will be an
enhanced intensity (c.f. Fig.3) or, interference increases the returning probability
and hence reduces the diffusion coefficient. This phenomenon is known as weak
localization.

     Since the displacement r scales with the mean free path, the width of the cone
of enhanced backscattering will increase and the diffusion constant will decrease
with increasing randomness of the medium. The width of the cone will be inversely
proportional to the mean free path.

     It is of tremendeous interest to study the enhanced backscattering in great
detail, both in theory and experiment, as there is an intimate connection between
enhanced backscattering and strong localization. Enhanced backscattering occuring
in the medium effectively reduces the diffusion constant. When enhanced
backscattering is so effective that the reduction is 100%, strong localization sets in.

2.2 Experimental Techniques

The first reports on enhanced backscattering as a result of interference were given
by Kuga and Ishimaru [6], the present authors [7] and by Wolf and Maret [8]. The
present authors [9] recorded cones of coherent backscattering using a set up of the
type drawn schematically in Fig.4. A linearly polarized laser beam was expanded
and then reflected from a beam splitter onto the sample. The intensity
back–scattered through the beamsplitter was recorded as a function of the
scattering angle, using a pinhole–detector assembly mounted on a stepper–motor
driven translation stage and positioned with the pinhole in the focal plane of the
lens L. In scans of high–viscosity samples the cell was spun around its axis to
average out the speckle that results from interference between different light paths.
Using a difference technique, the width and the enhacement factor of the
contribution to the backscatter cone were studied as a function of the depth z in

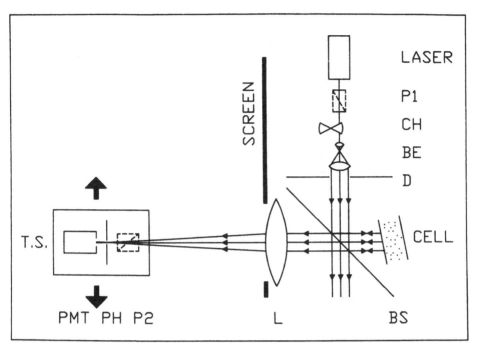

Fig.4   Experimental setup for the observation of backscattering enhancement.
           P1, P2, polarizers; CH, chopper; BE, beam expander; BS, beamsplitter;
           L, Lens; PH, pinhole; TS, translation stage.

the sample: subtracting the backscattering patterns of slabs of thickness L_1 and L_2
$(L_2>L_1)$, the scattering contributions coming from the front layer of the sample
$(z<L_1)$ cancel out, and what remains is the contribution of light that has "seen"
the deeper part $(L_1<z<L_2)$ of the slab (c.f. Fig.5).

      Etemad et al. [10] and Akkermans et al. [11] studied the effect of the cutting off
of long light paths both by limiting the sample thickness and by addition of
absorbing dye to the samples. In Etemad's work, a $\lambda/4$ plate was used in front of
the sample cell, which leads to elimination of a.o. the single–scattering
contribution to the backscattered intensity.

      In a study of solid bariumsulphate–air samples, Kaveh et al. [12] eliminated the
complication of polarization and angle–dependent transmission coefficients of the
beam splitter by using a mirror instead. In this way, the very top of the cone could
not be observed, but the "blind angle" was kept small by locating both the mirror
and the detector at a large distance from the sample.

      Quasi–2D samples of biological origin, and carefully constructed ones, were
studied using techniques similar to the ones described for 3–D by Freund et al. [13]
and van der Mark et al. [14] respectively.

      Vreeker et al. [15] studied the evolution in time of backscattering cones that
result from very short (100 fs) laser pulses, using a colliding–pulse mode–locked
(CPM) laser as a source and a light–gating detection technique based upon
second–harmonic generation.

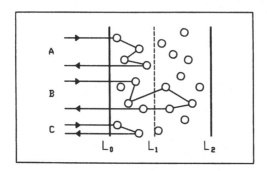

Fig.5   The intensity pattern of the "full slab" (L_0-L_2) contains contributions of
          A, B and C; that of the "front slab" (L_0-L_1) contains contributions of A
          and C. Subtraction of these patterns yields the pattern of the "deep
          slab" (L_1-L_2) containing only the "long path" contribution of B.

2.3 Weak Localization in 3–D Media

If linearly polarized light is incident upon a sample, then the polarization of the
scattered light may be partly or completely scrambled, so that the backscattered
light will consist of components with polarization parallel and perpendicular to
that of the incident light. Since light is a transverse vector wave, it always scatters
anisotropically. So in multiple scattering, the first few scattering events will result
in a transport of light through the sample that is not equally efficient in directions
parallel and perpendicular to the incident polarization vector. Thus the "cones" of
enhanced backscattering will not necessarily be cylinder–symmetrical. In the
backscattered intensity pattern one may look at the components with polarization
parallel and perpendicular to that of the incident beam (the parallel and
perpendicular cones). If in each case one records the intensity patterns while
moving the detector in directions parallel or perpendicular to that of the incident
polarization vector (parallel and perpendicular scans), a single sample will yield
four (in principle) non–equivalent intensity patterns.

2.3.1 Shape and Width of the Parallel Cone and Comparison to Scalar Theory

Cones of enhanced backscattering as recorded for optically thick samples while
monitoring the light component parallel to the incident polarization, show a
triangular top (the opening angle depends on the transport mean free path) and
slowly decaying wings (c.f. e.g. Fig.9). In Fig.6 (taken from Ref.16) the measured
width W of the cone in the parallel light component is plotted as a function of the
transport mean free path λ_{tr}. The expected inverse proportionality of W as a
function of λ_{tr} is seen to hold over the accessible range of nearly three decades.
     It may be verified that for the waves emerging from a time–reversed pair of
light paths, the components of the amplitude vectors parallel to the incident
polarization are equal, so that in the direction of pure backscattering interference
will be fully constructive. (For the perpendicular components this is not in general
the case). One could therefore try to describe the parallel cone using scalar theory.
Rigorous scalar theory for isotropic point scatterers, taking along ladder and

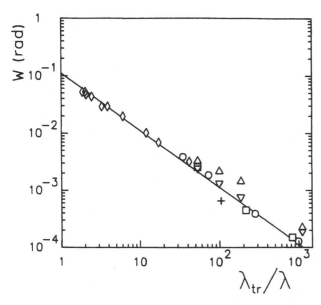

Fig.6   Full width W of the cone of enhanced backscattering vs. λ_{tr} for very thick slabs. $\lozenge$: 0.22 μm TiO$_2$ in 2–methylpentane–2.4–diol; $\circ$: 1.091 μm polystyrene speres in water; $\square$: 0.482 μm polystyrene spheres in water; $\triangle$: 0.214 μm polystyrene spheres in water ($\parallel$ scan); $\triangledown$: 0.214 μm polystyrene spheres in water ($\perp$ scan); +: 2.02 μm polyvinyltoluene spheres in water. Line: $W = (0.7/2\pi)\,(\lambda/\lambda_{tr})$, theoretical result for $L = \infty$.

cyclical terms, [16,17] predicts the width of the cone to relate with the transport mean free path λ_{tr} according to

$$W \simeq \frac{0.7}{2\pi}(\lambda/\lambda_{tr}).\qquad(3)$$

This relationship is compared with experimental data in Fig.6, and agreement is excellent.

The experimental enhancement factors $I_{top}/I_{background}$ depend somewhat on the nature of the sample (type of scatterers, concentration) and range from 1.5 to nearly 2. If only ladder and cyclical terms would contribute, any deviation of the enhancement factor from the value 2 should result from single scattering alone. (This simplest ladder term has no cyclical counterpart). It seems possible that other processes that either do not have a time–reversed counterpart (e.g. s_1,s_2,s_1 third–order) or give rise to interference terms that are angle–independent and therefore apparently contribute to the incoherent background s_1,s_2,s_3,s_1 fourth–order, etc.) contribute significantly in concentrated samples containing real (i.e. non–point–like) scatterers.

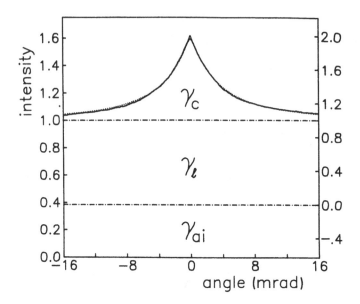

Fig.7   Enhanced backscattering of the parallel light component from a 2300 μm
            slab of a 1.05–vol% sample of $\phi \simeq$ 220 nm particles of TiO_2 in
            2–methylpentane–2,4–diol at λ_{vac}=514.5 nm. Solid line: experimental
            curve; dotted line: curve, calculated from diffusion theory, using $\lambda_{tr} =$
            9.7 μm. The left–hand vertical axis corresponds to the experimental
            curve, while the right–hand axis corresponds to the theoretical
            decomposition in contributions from the ladderterms γ_1, the interference
            terms γ_c, and the angle–independent terms γ_{ai}.

The numerical solution of rigorous scalar theory becomes rather
time–consuming if one wants to calculate the enhanced backscattering from thick
slabs. A versatile scalar diffusion approximation has also been developed [16,18].
Some results of this theory will be compared here to experimental data.

In Fig.7, the backscattering pattern as recorded using a thick slab ($L \simeq 240\lambda_{mf}$)
of 1.05 vol % and diameter $\phi \simeq 0.22$ μm rutile particles in
2–methylpentane–2,4–diol as a sample, is plotted together with the intensity
profile as calculated for this sample from diffusion theory. The λ_{tr} value used in
this calculation was obtained from λ_{sc} as determined from transmission
experiments [16] and $<\cos\theta>$ as found from Mie–theory assuming that the rutile
particles are monodisperse and spherical (which they are not). The theoretical
curve was fitted to the experimental data in the following way: the calculated
intensity profile was convolved with the instrumental resolution of 0.5 mrad. (The
advantage of convolving the calculated curve instead of deconvolving the
experimental one is that no assumptions regarding the shape of the latter are
needed). The tops of the resulting theoretical and experimental curves were then
superimposed and the vertical scale of the calculated curve was adapted so as to
make the outermost parts of its wings coincide with those of the experimental

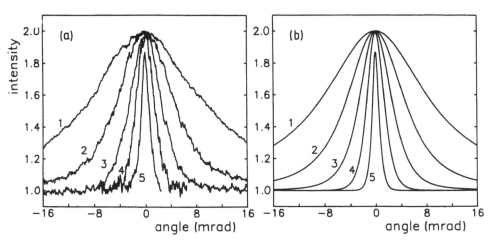

Fig.8   (a) Normalized (see text) difference cones recorded for the parallel light
         component using the sample of Fig.7 Difference slabs: curve 1: 0–13 μm;
         curve 2: 13–25 μm; curve 3: 25–50 μm; curve 4: 50–100 μm; curve 5:
         100–1000μm. (b) Normalized backscattering curves, calculated from
         diffusion theory and convolved with the resolving power of 0.5 mrad for
         the difference slabs of (a), using $\lambda_{tr} = 9.7$ μm. These figures illustrate
         that the deeper a difference slab is situated in the sample, the narrower
         its contribution to the cone (as expected, because light paths visiting
         deep slabs are necessarily long).

cone. The shape and width of the calculated cone is found to fit perfectly to the
experimental data. At the same time we see that the observed intensity profile has
an offset with respect to the calculated one that amounts to approximately 30 % of
the total background. In Fig.7, this angle–independent contribution is denoted by
γ_{ai}.

Fig.8a shows so called "difference cones" that were obtained subtracting the
intensity patterns of slabs of different thicknesses, using the sample that gave the
full cone of Fig.7. In order to present several curves in the same figure without
loosing clarity, a normalization procedure was performed that is more conveniently
outlined if we first discuss the corresponding calculated cones. The latter are
plotted in Fig.8b and were obtained in the following way: the intensity profiles
were calculated from diffusion theory, and then convolved with the instrumental
resolution. The resulting (convolved) curves were normalized, each one with
respect to its own background level. We return now to the experimental curves:
the vertical scales of the cones in Fig.8a were adapted so as to match their heights
(intensities) to those of their calculated counterparts. We conclude that the width
and shape of the parallel cone are correctly described by scalar theory.

2.3.2 Vector Character Effects and their Interpretation

In addition to the enhancement in the parallel light component, a less pronounced
enhancement (enhancement factors typically $1.0 - 1.3$) is found in the

perpendicularly polarized component. The appearance of both the parallel and the perpendicular cone depends on a.o. the nature of the scattering medium. It appears that we can distinguish essentially two types of scatterers, the small point–dipole scatterer (Rayleigh scatterer) and the bigger "spotlight" scatterer, which scatters in the forward direction mainly. Intensity profiles as recorded for both the parallel and perpendicularly polarized components of light, backscattered from 9.6 vol% of 0.214 and 1.092 μm polystyrene microspheres are shown in Fig.9a and 9b respectively. The following differences may be noted:

* For the smaller scatterers, polarization is partly retained in the backscattered light (as may be seen from the relative levels of the incoherent background for both polarizations) indicating that (very) low–order processes play an important part in the total backscattering. For the larger scatterers, the scambling of polarization is very nearly complete.
* For the smaller scatterers the parallel cone is spatially anisotropic, being broader in the scan parallel to the polarization of the incident beam than in the perpendicular direction. The triangular tops are equally wide in both scans, indicating that the broadening of the cone in the parallel scan is a lower–order multiple scattering effect.
* The perpendicular cones have rounded–off tops, indicating that only short paths (lower–order scattering processes) contribute.

A diffusion approximation for point–dipole scattering has been developed [19] which predicts the slight enhancement in the perpendicular component, but not the anisotropy in the parallel cone. Assuming point–dipole scattering it may be shown [20] that:

* In the backward direction the ratio in which the parallel and perpendicular components contribute to the incoherent background is 8 for second–order scattering and for higher orders gradually converges to 1.
* The perpendicular enhancement factor is 2 for second order backscattering and gradually converges to 1 for higher orders.
* Backscattering via second–order processes proceeds more efficiently along "light paths" lying in or near a plane perpendicular to the incident polarization vector, than along light paths in or near a plane including this vector. A displacement of the detector in a direction parallel to the incident polarization vector will therefore influence the relative phases of the more strongly contributing time–reversed pairs to a lesser extent (and hence reveal a broader cone) than one in the perpendicular direction.

We conclude that all vector effects observed for samples containing the smaller scatterers may be explained in terms of the large relative importance of lower–order multiple scattering processes in the Rayleigh case.

The larger micro–spheres scatter mainly in the forward direction ("spotlight") and as a result, very–low–order processes do not contribute significantly. Still, there is a (rounded–off) perpendicular cone, indicating that for the shortest significantly contributing lightpaths the perpendicular components in the time–reversed pairs interfere constructively. Figs.10a en 10b (taken from Ref. 21) compare the evolution of polarization by the shortest significantly contributing

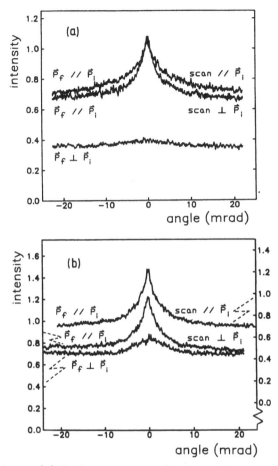

Fig.9   (a) Backscatter cones from a 2400 μm slab of 9.6 volume % of 0.215 μm
        polystyrene spheres in water. Upper curve: spatial scan ‖ to
        polarization vector of polarized component; middle curve: spatial scan ⊥
        to polarization vector of polarized component. The two curves differ
        approximately by a factor of two in width. Lower curve: depolarized
        component (only one spatial scan is shown as no anisotropy was
        detected). (b) Same as (a) but for a 1500 μm slab of 9.6 vol. % of 1.091
        μm polystyrene particles.

lightpaths in the small– and large–particle regime respectively. It is seen that for
both regimes these paths lead to constructive interference in the perpendicular
component. In the large–particle regime these paths completely scramble
polarization, and are equally efficient whether parallel or perpendicular to the
incident polarization vector, which explains the absence of spatial anisotropy of the
cones for this regime.

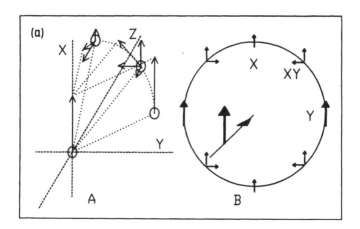

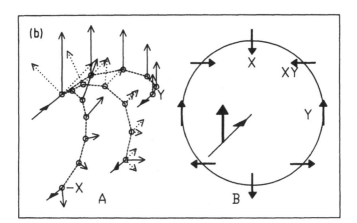

Fig.10   (a) A: Three second–order light paths for Rayleigh scattering;
          B: Direction and amplitude of backscattered wave in second order
          Rayleigh scattering as a function of the relative orientation of the
          scattering particles. (b) A: Three "lowest–order" light paths in (large
          particle) Mie scattering; B: Direction and amplitude of the
          backscattered wave as a function of the orientation of the "shortest"
          light path in Mie scattering.

2.3.3 Time–Resolved Coherent Backscattering

The relationship between path length and width of the contribution to the cone
was studied in a time–resolved experiment. Fig.11 shows cones recorded at 30 fs
and 330 fs after the CPM–laser pulse respectively, using a 6.8 vol%
TiO_2/2–methylpentane–2,4–diol sample $(\lambda_{tr} \simeq 2\ \mu m)$. The theoretical curves were
calculated as follows:

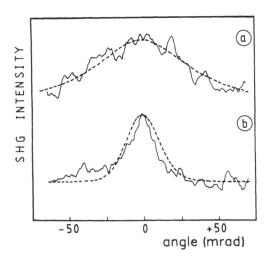

Fig.11　　Enhanced backscattering cones measured for a 6.8 vol% suspension of TiO$_2$ particles in 2–methylpentane–2.4–diol ($\lambda_{mf} = 2$ μm) at (a) 30 fs and (b) 330 fs after the laser pulse. Dashed curves represent results of calculations based on (6).

The approximate diffusion theory developed by Akkermans et $al.$ [18] in which the shape of the backscattering cone is found by integration over all contributing path lengths, neglecting the depth of the first and last scatterer, was used to find the time dependence of the coherent intensity $\gamma(\theta,t)$:

$$\gamma(\theta,t) \propto t^{-3/2}[1+\exp\{-4\pi^2 ct\lambda_{mf}\theta^2/(3\lambda^2)\}], \tag{5}$$

where c is the speed of light. The equation describes a Gaussian shaped backscattering cone with a width of the order $\lambda/(\lambda_{mf}ct)^{1/2}$. In principle the equation is only valid for times t long enough for the scattering to have reached the diffusive limit, but this condition was apparently met within the error of the experiment, since the data are well described by the theory. The observed intensity at a particular angle θ still has to be convolved with the pulse shape. We then get

$$\tilde{\gamma}(\theta,t) = \iint \gamma(\theta,\tau-t)\, I_p(t)\, I_p(t'-\tau)\,dtdt', \tag{6}$$

in which the first integral gives the backscattered signal as a convolution of the impulse response function of the scattered $\gamma(\theta,t)$, and the incident pulse shape $I_p(t)$. The second integral gives the measured signal as a convolution of the backscattered signal with a lightgating pulse, which is used in case of ultra fast detection.

In Fig.11 theory and experiment are compared. The structure in the experimental curves is due to speckle, which in a time–resolved experiment on femtosecond scale can not be averaged out by e.g. spinning the sample for practical reasons. The agreement between theory and experiment is excellent.

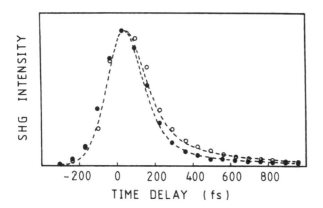

Fig.12   Backscattered light pulses from the sample of Fig.11 measured for $\theta = 0$
            mrad (open circles) and $\theta = 27$ mrad (full circles). Dashed curves
            represent results of calculations based on (6).

Fig.12 shows the time decay of pulses scattered from the sample, as measured
at fixed angles with the direction of pure backscattering. At exact backscattering
interference between time–reversed pairs is constructive at all times. At small
angles with the backscattering direction, the time–reversed pairs dephase with $t^{1/2}$.
and hence their coherent contribution decays during a short period following the
laser flash. At longer times, only the incoherent contribution persists, which
corresponds to half the intensity at pure backscattering.

3. ON THE ROAD TOWARDS STRONG LOCALIZATION

3.1 Very Strongly–Scattering Media and Localization

The occurrence of weak localization being well established, the question as to
whether it will be possible to observe strong localization of light should be
considered. The crucial parameter is the transport mean free path λ_{tr}, which
should be as short as possible.

It seems that a suspension of TiO_2 in air or in some low–refractive–index
substance is the system in which the shortest λ_{tr}–values for visible light may be
realized in practice. Once $\lambda_{tr} < \lambda_{cr}$ (where λ_{cr} is some critical mean free path) the
system will be localized. The Ioffe–Regel criterion gives λ_{cr} as $\lambda/(2\pi)$, and indeed
the existence of delocalized states at $\lambda_{tr} < \lambda/(2\pi)$ seems to be very unlikely. On
the other hand it can not be excluded beforehand that the localization transition
might occur earlier. To detect strong localization one should study light that
traveled through the medium over at least the localization length, λ_{loc}, which near
the localization transition may be larger than λ_{tr} by orders of magnitude. Since in
backscattering lower–order scattering contributions are dominant, transmission
experiments seem to be indicated for the detection of strong localization. Whereas
in the regime of classical diffusion the transmission decays with increasing length L
of the sample according to L^{-1}, in the localized regime this decay is expected to be
faster: possibly starting as L^{-2} for $L < \lambda_{loc}$ and exponential for $L > \lambda_{loc}$.

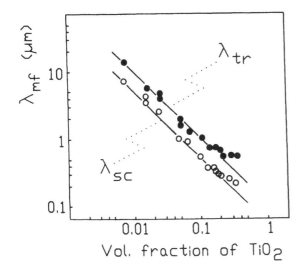

Fig.13   Scattering mean free path (lower plot) and transport mean free path
          (upper plot) as a function of the volume fraction of $\phi \simeq 220$ nm TiO_2
          particles.

Absorption by the sample is a complicating factor, as it also leads to a faster than
L^{-1} decay.

Watson *et al.* [22] performed time–resolved experiments on the diffusion of
light through scattering media. For polystyrene–sphere suspensions in water with
$2\pi\lambda_{mf}/\lambda \simeq 360$, they found no signs of departure from classical diffusive behavior.
In TiO_2–water and TiO_2–air samples, they could not determine λ_{mf}, because of
absorption.

The present authors determined λ_{sc} and λ_{tr} for TiO_2 suspensions in
2–methylpentane–2,4–diol. Results are shown in Fig.13. It is seen that, upon
increasing the volume fraction of TiO_2 above $\simeq 15\%$, the transport mean free path
seems to saturate at a value that is slightly larger than the wavelength of the light
in the medium. (The width of the cone of enhanced backscattering equally
saturates.) It is difficult to tell whether particle–particle correlations, dependent
scattering or impedance mismatch at the boundaries is the main cause of this
effect, but since the average inter–particle separation at this volume fraction is
roughly equal to the wavelength in the medium, dependent scattering may be
expected to at least play a part. Thus, in the samples with strongest randomness,
λ_{tr} is still by a factor of 8 higher than the Ioffe–Regel value for λ_{cr}. A systematic
study of the diffusion coefficient as a function of the sample–thickness might reveal
how near these samples come to the localization transition.

3.2.   The Influence of Impedance Mismatch at the Boundaries

In all the models up to now the following restrictive assumption on the boundary
condition was adopted: when the light diffuses to the boundary it will escape with
probability one. This allows one to use the simplifying reflection principle in the

calculation of Green's functions [23]. The experimental optimization of the multiple scattering in these types of materials requires the maximization of the contrast in index of refraction in the medium. This results usually in a situation in which the average index of refraction in the medium differs considerably from the index of refraction of the outside world thus causing internal (and external) reflection. Our estimate is that in strongly scattering media the internal reflection coefficient, averaged over all angles, could be as high as 0.85. This urges the question to what extent does this finite reflectivity influence the dynamics of the propagation. We have recently developed a theory which incorporates the effect of internal reflection [24]. For experiments requiring backscattering geometries, like coherent backscattering and time–resolved incoherent backscattering it was demonstrated that internal reflection seriously modifies the outcome. It turns out that the backscattering cone becomes narrower when one introduces a finite internal reflection. The fact that the cone becomes narrower on increasing the reflectivity can easily be understood. The influence of finite reflectivity is to force the light paths in the sample to be longer. But longer light paths get out of phase easier and consequently the backscattering cone becomes narrower.

In transmission experiments the significance of the corrections depends on the thickness of the slab (in units of mean free path) and the influence disappears for very thick slabs. To a large extent this effect can be accounted for by renormalizing the diffusion coefficient with a length–scale dependent reduction factor. Of course the overall absolute intensities are always seriously affected by finite reflection coefficients.

The occurrence of internal reflection can be circumvented by matching of index of refraction. We think that the experiments performed in the past where the coherent backscattering has been measured for titania samples with air as medium could have been seriously affected by the effect of internal reflection. The interpretation of experiments with pulses in backscattering and transmission through thin slabs will seriously be influenced by the presence of internal reflection. Also in other experiments where reflection geometries are used or where transmission through relatively thin samples is studied, including fluctuation experiments, the influence of internal reflection could be sizeable.

3.3 Direct Measurement of the Diffusion Constant

Since strong localization implies the absence of diffusion, measurement of the diffusion constant is the most direct way to detect localization, Genack et al. [25] determined the diffusion constant for light in TiO_2–polystyrene and in 70 vol% sintered TiO_2 samples from the rate of change of the speckle pattern observed in transmission on changing the wavelength of the light. They found diffusion constants corresponding to mean free paths of 1.4 μm and 350 nm respectively.

The present authors used Genacks technique on TiO_2–air samples of a lower volume fraction. From the width at half height $\Delta\nu_{1/2}$ of the correlation function $<I(\nu)I(\nu+\Delta\nu)>$ the diffusion constant D was determined to be $\simeq 12$ m^{2}s^{-1}, and estimating the velocity of light in the sample to be $\simeq 1.7\times 10^8$ ms^{-1}, we obtain a value of $\simeq 200$ nm for λ_{tr}. This is indeed considerably less than the $\simeq 350$ nm value, found from the width of the backscatter cone, and the difference is thought to be due to the effect of internal reflection at the interface.

Genack et al. [26] also determined diffusion constants using time–resolved techniques. Very recently [27] they found a diffusion constant as small as 1.45 m^{2}s^{-1} (corresponding to a mean free path of $\simeq 25$ nm) in samples of close–packed

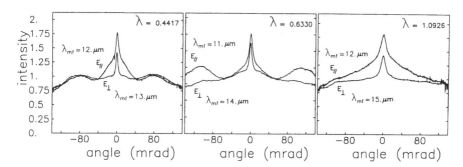

Fig.14   Backscattered intensity of our sample as a function of scattering angle.
The transport mean free path for both $E_{\parallel}$ and $E_{\perp}$ are given for three
different wavelengths (in μm).

titania spheres. Since the experimental value for the mean free path is expected to
correspond to the difference $\lambda_{tr}-\lambda_{cr}$ [28] with λ_{cr} the critical length for
localization, their samples might indeed be very close to the localization transition.

3.4   Optical Localization in a Two Dimensional System (2–D)

Up till now it has not been possible to find strong localization in a 3–D sample.
The prediction is, that in an infinite 2–D system, any degree of disorder will lead
to localization, while in 3–D a certain critical degree of disorder is needed. In a
2–D system, even if it is finite or suffers from some loss, it might be possible to
obtain Anderson localization.

In order to realize a 2–D medium, we constructed a 25 mm wide sample of 40
cm long glass fibres with a diameter of 5 μm and an estimated parallellism of 0.2
mrad. The thickness of the sample was approximately 70 μm. The sample was
illuminated (at right angle) with 20 mm wide laser beams with wavelengths λ of
1092.6 nm, 633 nm and 441.7 nm respectively. Both the transmitted and
backscattered light were imaged onto an optical multichannel analyzer. In Fig.14,
the backscattered intensity for both parallel ($E_{\parallel}$) and perpendicular ($E_{\perp}$)
polarization of the incident and detected light with respect to the fibre direction as
a function of scattering angle (in the 2–D plane) is shown. At precisely
backscattering we find a sharp peak of which the width is dependent on wavelength
and polarization, and is the result of weak localization. The structure in Fig.14 is
the consequence of the single–fibre scattering properties. The transport mean free
path λ_{mf} can be estimated from the width of the sharp peak or calculated with
Mie–theory; the values are consistent within the experimental error, and are given
in Fig.14.

The degree of polarization of the backscattered intensity was measured to be
better than 600:1, limited by the quality of the polarizer used. This ratio is a
measure for the dimensionality ($d=2+\epsilon$) of the system, and should be infinite for a
truly 2–D system ($\epsilon=0$). Another measure for the dimensionality is the amount of
light which is scattered out of the 2–D plane. In Fig.15 the backscattered intensity
of our sample is shown as a function of angle perpendicular to the 2–D plane. The
width of the peak is a measure for ϵ by approximately $\epsilon \simeq 2^{1/2}\times$ Width. Using the

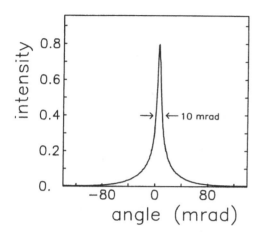

Fig.15   The backscattered intensity of our sample as a function of the angle
          perpendicular to the 2–D plane. The width of the peak is a measure for
          ϵ, and was found to be virtually independent of wavelength and
          polarization.

criterion given by John [29]: $(2\pi\lambda_{mf}/\lambda)^{1+\epsilon} <\sim \epsilon^{-1}$, we find that we for the first time
have entered the localization regime for $\lambda = 1092.6$ nm and $E_{\parallel}$. Since the
localization length probably exceeds the size of the sample, only the so–called
logarithmic correction [30] to the amount of transmitted light might be observed
here. We are performing transmission experiments on samples with different sizes
now.

ACKNOWLEDGMENT

This work is part of the research program of the "Stichting voor Fundamenteel
Onderzoek der Materie" (FOM), which is financially supported by the
"Nederlandse Organisatie voor Wetenschappelijk Onderzoek" (NWO).

REFERENCES

[1]   P.W. Anderson, Phys. Rev. 109 (1958) 1492.
[2]   For a recent collection of papers on Anderson localization, see: *Anderson
       Localization*, eds. T. Ando and H. Fukuyama (Springer, Berlin, 1988).
[3]   W. Götze, J. Phys. C12 (1979) 1279; W. Götze, Phil. Mag. B43 (1981) 219.
[4]   D. Vollhardt and P. Wölfle, Phys. Rev. B22 (1980) 4666.
[5]   N.F. Mott, Adv. Phys. 16 (1967) 49; see also: N. Gee and G.R. Freeman,
       Can. J. Chem. 64 (1986) 1810.
[6]   Y. Kuga and A. Ishimaru, J. Opt. Soc. Am. A1 (1984) 831.
[7]   M.P. van Albada and A. Lagendijk, Phys. Rev. Lett. 55 (1985) 2692.
[8]   P.E. Wolf and G. Maret, Phys. Rev. Lett. 55 (1985) 2696.

[9] M.P. van Albada, M.B. van der Mark and A. Lagendijk, Phys. Rev. Lett. 58 (1987) 361.
[10] S. Etemad, R. Thompson, M.J. Andrejco, S. John, and F.C. MacKintosh, Phys. Rev. Lett. 59 (1987) 1420.
[11] P.E. Wolf, G. Maret, E. Akkermans and R. Maynard, J. Phys. France 49 (1988) 63.
[12] M. Kaveh, M. Rosenbluh, I. Edrei and I. Freund, Phys. Rev. Lett. 57 (1986) 2049.
[13] I. Freund, M. Rosenbluh, R. Berkovits, and M. Kaveh, Phys. Rev. Lett. 61 (1988) 1214.
[14] M.B. van der Mark and A. Lagendijk, EQEC '88 (1988); M.B. van der Mark and A. Lagendijk, QELS '89 (1989).
[15] R. Vreeker, M.P. van Albada, R. Sprik, and A. Lagendijk, Phys. Lett. A132 (1988) 51.
[16] M.B. van der Mark, M.P. van Albada, and A. Lagendijk, Phys. Rev. B37 (1988) 3575.
[17] L. Tsang and A. Ishimaru, J. Opt. Soc. Am. A1 (1984) 836; ibid. 2 (1985) 1331; ibid. 2 (1985) 2187.
[18] E. Akkermans, P.E. Wolf, and R. Maynard, Phys. Rev. Lett. 56 (1986) 1471.
[19] M.J. Stephen and G. Cwilich, Phys. Rev. B34 (1986) 7564; F.C. MacKintosh and S. John, Phys. Rev. B37 (1988) 1884.
[20] M.P. van Albada and A. Lagendijk, Phys. Rev. B36 (1987) 2353.
[21] M.P. van Albada, M.B. van der Mark and A. Lagendijk, J. Phys. D21 (1988) S28.
[22] G.H. Watson, Jr., P.A. Fleury, and S.L. McCall, Phys. Rev. Lett. 58 (1987) 945.
[23] P.M. Morse and H. Feshbach, Methods of Theoretical Physics (McGraw–Hill, New York, 1953).
[24] A. Lagendijk, R. Vreeker and P. de Vries, Phys. Lett. A136 (1989) 81.
[25] A.Z. Genack, Phys. Rev. Lett. 58 (1987) 2043.
[26] A.Z. Genack, L.A. Ferrari, J. Zhu, N. Garcia, J.M. Drake, Optical and Microwave Propagation in Random Dielectric and Metallic Media, proceedings of the 1987 ILS conference (1987).
[27] J.M. Drake and A.Z. Genack, Phys. Rev. Lett. 63 (1989) 250.
[28] P.W. Anderson, Philos. Mag. B52 (1985) 505.
[29] S. John, Phys. Rev. B31 (1985) 304.
[30] R.A. Davies and M. Pepper, J. Phys. C 15 (1982) L371.

M.P. van Albada, A. Lagendijk and M.B. van der Mark are with the Natuurkundig Laboratorium der Universiteit van Amsterdam, Valckenierstraat 65, 1018 XE Amsterdam, The Netherlands.   A. Lagendijk is also with the FOM–Instituut voor Atoom– en Molecuulfysica, Kruislaan 407, 1098 SJ Amsterdam, The Netherlands.

PHOTON LOCALIZATION: THE INHIBITION OF ELECTROMAGNETISM IN CERTAIN DIELECTRICS

Sajeev John

We discuss the properties of a new class of dielectric materials which exhibit strong localization of photons in analogy to the localization of electrons in disordered semiconductors. These materials consist of lossless high dielectric contrast scattering microstructures with short range spatial correlation. The existence of short range order leads to a fundamental modification of the traditional Ioffe–Regel criterion for localization making photon localization experimentally observable in three dimensions.

1. INTRODUCTION

The semiconductor is a material of fundamental importance to electronically based technology. The essence of its central role in electronics is that it is a material in which the spectrum or energy–momentum relationship of electrons can be tailored by the coherent scattering of the electron from a periodic array of atoms. The structure and composition of the materials determine the positions of allowed energy bands and forbidden gaps. A light wave, by analogy, may also scatter coherently from dielectric microstructures. Under very specific conditions, the dielectric scatterers can play a role analogous to the atoms of a semiconductor. The electromagnetic spectrum may, likewise, be tailored by the structure and composition of the photonic material. Since the electromagnetic force is the fundamental force of condensed matter physics, such an alteration of the electromagnetic force is likely to have important consequences. It is the purpose of this article to elucidate the special conditions on structure and composition which such a photonic material must satisfy. By analogy with semiconductors, one of the most apparent consequences is the localization of light [1,2].

Unlike the band tail localization of electrons in a random potential, for the case of light, localization occurs in the positive energy continuum of electromagnetic states at energies higher than the highest potential barrier. However, in both optics and electronics the concept of short range order in the scattering microstructures is central. This fundamental concept provides an essential alternative to the traditional Ioffe–Regel criterion for photon localization which, as I will argue, guarantees the experimental observability of photon localization in a higly controlled and predictable manner in three dimensional disordered dielectrics.

2. ELECTRONIC LOCALIZATION AND THE OPTICAL ABSORPTION EDGE IN SEMICONDUCTORS

Electronic localization in semiconductors provides a valuable paradigm for the physics of photon localization in nonabsorbing dielectrics. The spectrum of strongly localized electronic states in disordered semiconductors may be probed by

W. van Haeringen and D. Lenstra (eds.), Analogies in Optics and Micro Electronics, 105–115.

optically induced electronic transitions from the filled valence band to empty conduction band tail states. Although the existence of subgap optical absorption due to localized electronic states has long been recognized, a quantitative theory of the observed absorption edge, which is physically transparent, has remained elusive until recently. The key concept, as I will describe, is that electronic localization in the pseudogap of an amorphous semiconductor is not merely a consequence of disorder but rather an interplay between order and disorder, which I shall refer to as short range order.

In 1953, Urbach [3] proposed the empirical rule for the optical absorption coefficient $\alpha(\omega)$ associated with electronic transitions from the valence to conduction band tail in a disordered solid. As originally applied to silver- and alkali-halides, this rule states that for $\omega \lesssim \omega_0$,

$$\alpha(\omega) \propto \exp[\hbar\omega - \hbar\omega_0)/E_0],\tag{1}$$

where $\hbar\omega$ is the photon energy and E_0 and $\hbar\omega_0$ are fitting parameters, E_0 being proportional to kT in Urbach's original work. Subsequent studies [2] on disordered semiconductors and glasses exhibiting this Urbach exponential spectral behavior have strongly suggested that the Urbach absorption edge is a nearly universal property of disordered solids attributable to exponential band tails and that the underlying physics is both simple and general.

It is reasonable to relate the spectral dependence (1) with an associated exponential tail in the density of states for an electron in a static random potential [4,5]:

$$[\frac{-\hbar^2 \nabla^2}{2m} + V(x)]\,\psi(x) = E\psi(x).\tag{2}$$

Here $<V(x)> = 0$ and we are interested in energies $E < 0$. Using the central limit theorem, the various forms of disorder contribute to an essentially gaussian probability distribution for the Fourier components $V(k)$ of the potential:

$$P\{V(k)\} \propto \exp[-S\{V\}], \quad S = \tfrac{1}{2}\int \frac{d^d k}{(2\pi)^d}\, V(k)B^{-1}(k)V(-k)\tag{3a}$$

and for convenience the autocorrelation function

$$B(k) = (V_{\rm rms})^2(\pi L^2)^{d/2}\exp(-k^2L^2/4)\tag{3b}$$

is characterized by a correlation length L measuring the spatial extent of short range order, typically of order the interatomic spacing. Band tail states arise in such a model from potential fluctuations of depth V_0 large compared to the typical functuation $V_{\rm rms}$, and the corresponding probability of occurrence is exponentially small. For instance, if we consider a potential fluctuation of the form

$$V(x) = -V_0\exp(-x^2/a^2),\tag{4}$$

the probability of occurrence is determined by

$$S = \frac{(V_0)^2}{2(V_{rms})^2} \left[\frac{a}{L}\right]^d \left[2-\left[\frac{L}{a}\right]^2\right]^{-d/2}. \qquad (5)$$

This defines a variational problem for the class of gaussian potentials parametrized by a depth V_0 and a range a. The requirement that such a potential fluctuation contributes to the electronic density of states at an energy $-|E|$ places a constraint on the variation parameters V_0 and a. For $|E| >> V_{rms}$ this becomes the requirement that the potential (4) possesses a ground state at precisely an energy $-|E|$ since higher order bound states at that energy would require a considerably less probable potential.

In terms of the energy scale $\epsilon_L \equiv \hbar^2/(2mL^2)$ the resulting density of states exhibits an essentially linear exponential behavior

$$\rho(E) \sim e^{-S} \sim \exp\left[-14.4 \frac{|E| \epsilon_L}{(V_{rms})^2}\right] \quad (d=3) \qquad (6)$$

over a range $0.1 < |E|/\epsilon_L < 2$. For $\epsilon_L \simeq 0.5$ eV, this occupies most of the experimentally relevant energy range. For $|E|/\epsilon_L >> 2$, this crosses over to the gaussian density of states whereas for $|E|/\epsilon_L << 0.1$ there is a Halperin–Lax [6] density of states for which $\ln \rho(E) \sim -7.0 |E|^{1/2} (\epsilon_L)^{3/2}/(V_{rms})^2$. The fact that the crossover regime dominates the entire experimentally observabe region of the band tail is a direct consequence of short range order.

At energies higher than the strongly localized regime, the simple potential well picture becomes less applicable. A critical point is eventually reached at which a transition to extended electronic states occurs. This transition is well described by the scaling theory of localization [7] which pays no specific attention to the details of short range order in the solid. This is depicted in Fig.1. For weak disorder there is a tail of strongly localized states, referred to as the Urbach edge, for $E<0$ which is separated by a mobility edge from positive energy extended states in the squareroot continuum. As the disorder parameter V_{rms} is increased the mobility edge eventually moves into the positive energy continuum. The detailed trajectory of the mobility edge as a function of disorder has been the subject of extensive numerical studies [8]. The noteworthy point, however, for present considerations is that only in the limit of very strong disorder do states with $E > 0$ exhibit localization.

3. PHOTON LOCALIZATION IN DISORDERED DIELECTRICS

In the case of monochromatic electromagnetic waves of frequency ω propagating in a disordered nondissipative dielectric medium with dielectric constant $\epsilon(x) = \epsilon_0 + \epsilon_{fluct}(x)$, the wave equation for the electric field vector E may be written in a form resembling the Schrödinger equation:

$$-\nabla^2 E + \nabla(\nabla \cdot E) - \frac{\omega^2}{c^2} \epsilon_{fluct}(x)E = \epsilon_0 \frac{\omega^2}{c^2} E. \qquad (7)$$

Here the randomly fluctuating part of the dielectric constant $\epsilon_{fluct}(x)$ plays the role of the random potential $V(x)$ with a mean value of zero and some appropriate spatial autocorrelation function. For nonmetallic scatterers with an everywhere

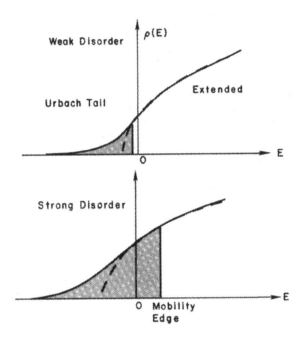

Fig.1   Upper part: one electron density of states in a correlated Gaussian
         random potential. For weak disorder there is an Urbach band tail of
         localized states below the positive energy square root continuum of
         extended states which is shifted downward slightly by the disorder.
         Lower part: as the disorder is increased, the mobility edge eventually
         moves into the positive energy regime.

real positive dielectric constant, it is evident that the energy eigenvalue $\epsilon_0 \omega^2 / c^2$ is
always positive thereby precluding the possibility of band tail localization below
the continuum. Herein lies the fundamental challenge of photon localization: in
traditional electronic systems localization occurs at low energies in which the
electron is trapped in potential wells or faces a number of large barriers. In the
case of classical waves, however, as the energy eigenvalue is lowered by going to
lower frequency ω, the strength of the scattering potential $(\omega^2/c^2)\epsilon_{fluct}(x)$ also
vanishes and the states are extended. In the opposite limit of very high frequencies,
states are again extended irrespective of the strength of the disorder. Unlike the
familiar picture of electronic localization, what we are really seeking in the case of
light is an intermediate frequency window of localization within the positive energy
continuum! To further complicate matters, the condition that $\epsilon(x) > 0$ requires
that the energy eigenvalue $\epsilon_0 \omega^2 / c^2$ is greater than the highest potential barrier
given by $(\omega^2/c^2)|\epsilon_{fluct}(x)|$.
     The underlying physics of the high and low frequency limits in the case of
photons which have no upper frequency cuttoff can be made more precise by
considering scattering from a single dielectric sphere. Consider a plane wave of

wavelength λ impinging on a small dielectric sphere of radius $a << \lambda$ of dielectric constant ϵ_a embedded in a uniform background dielectric ϵ_b in d–spatial dimensions. The scattered intensity I_{scatt} at a distance R from the sphere can be a function of only the incident intensity I_0, the dielectric constants ϵ_a and ϵ_b and the lengths R, λ and a. In particular I_{scatt} must be proportional to the square of the induced dipole moment of the sphere which scales as the square of its volume $\sim(a^d)^2$ and by conservation of energy must fall off as $1/R^{d-1}$ with distance from the scattering center:

$$I_{scatt} = f_1(\lambda,\epsilon_a,\epsilon_b) \; \frac{a^{2d}}{R^{d-1}} \; I_0. \tag{8}$$

Since the ratio I_{scatt}/I_0 is dimensionless, it follows that $f_1(\lambda,\epsilon_a,\epsilon_b) = f_2(\epsilon_a,\epsilon_b)/\lambda^{d+1}$ where f_2 is another dimensionless function of the dielectric constants. The vanishing of the scattering cross section for long wavelengths as $\lambda^{-(d+1)}$ is the familiar result for why the sky is blue in three dimensions. This generalization of Rayleigh scattering to d–dimensions also reveals the origin of diverging localization lengths for one and two–dimensional classical waves and of course the existence of extended states in $d = 3$. For a dense random collection of scatterers this behavior remains evident in the elastic mean free path l which is proportional to λ^{d+1} for long wavelengths [9]. In one and two dimensions, all states are localized but with diverging localization lengths ξ_{loc} behaving as $\xi_{loc} \sim l$ in one dimension and $\xi_{loc} \sim l \exp(\omega l/c)$ in two dimensions [9].

It is likewise instructive to consider the opposite limit in which the wavelength of light is small compared to the scale of the scattering structures. For scattering from a single sphere it is well known [10] that for $\lambda << a$ the cross section saturates at a value $2\pi a^2$. This is the result of geometric optics. For a dense random collection of scatterers it is useful to introduce the notion of a correlation length a. On scales shorter than a, the dielectric constant does not vary appreciably and wave propagation is well described in WKB approximation. The essential point is that the elastic mean free path never becomes smaller than the correlation length. If one adopts the most naive version of the Ioffe–Regel [11] condition $2\pi l/\lambda \simeq 1$ for localization, with λ being the vacuum wavelength or even an effective medium wavelength of light, it follows that extended states are expected at both high and low frequencies. However as depicted in Fig.2, for strong scattering, there arises the distinct possibility of localization within a narrow frequency window when the quantity $\lambda/2\pi \simeq a$. It is this intermediate frequency regime which we wish to analyze in greater detail.

For point like scatterers, a straightforward first principles calculation [12] in $d = 2 + \epsilon$ dimensions yields the criterion $(l\omega/c)^{d-1} \simeq 1/\epsilon$ for the occurrence of a mobility edge. This result is asymptotically exact as $\epsilon \to 0$. Extrapolating to $\epsilon = 1$ yields in three dimensions what I will refer to as the free–photon Ioffe–Regel condition. This particular result is based on perturbation theory about free photon states which undergo multiple scattering from point–like objects and a disorder average is performed over all possible positions of the scatterers. In effect, statistical weight is evenly distributed over all possible configurations of the scatterers and the medium has an essentially flat structure factor on average. The first correction to this picture is to associate some nontrivial structure to the individual scatterers. The theory of Mie resonances for scattering from dielectric spheres immediately tells us that this can have profound consequences on the elastic mean free path. For instance for spheres of radius a, of dielectric constant ϵ_a

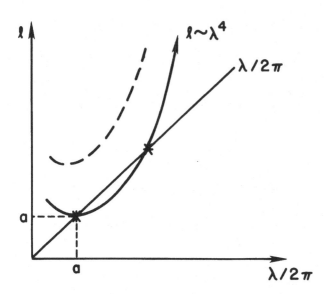

Fig.2   Behavior of the elastic mean free path as a function of wavelength. In
         the long wavelength Rayleigh scattering limit $l \sim \lambda^4$. In the short
         wavelength limit $l \gtrsim a$, the correlation length. For a strongly disordered
         medium (solid curve) there may exist a range of wavelengths for which
         $2\pi l/\lambda \simeq 1$ exhibiting localization. This would not occur for dilute
         scatterers (dashed curve).

embedded in a background dielectric ϵ_b, and for a ratio $\epsilon_a/\epsilon_b \simeq 4$, the first Mie
resonance which occurs at a frequency given by $(\omega/c)2a \simeq 1$, yields a scattering
cross section $\sigma \simeq 6\pi a^2$. For a relatively dilute collection of spheres of number
density n, the classical elastic mean free path becomes $l \sim 1/(n\sigma) = 2a/(9f)$. Here I
have introduced the volume filling fraction f of the spheres. Extrapolating this
dilute scattering result to higher density, it is apparent that for a filling fraction
$f \simeq 1/9$, the free–photon Ioffe–Regel condition is satisfied on resonance. It is
tempting to increase the density of scatterers so as to further decrease the mean
free path. However, the fact that the cross section on resonance is 6 times the
geometrical cross section indicates that a given sphere disturbs the wavefield over
distances considerably larger than the actual sphere radius. The existence of the
resonance requires that the "spheres of influence" of the scatterers do not overlap.
Indeed for higher densities the reader may verify that the spheres become optically
connected in this sense and that the mean free path in fact increases rather than
decreasing. From the single scattering or microscopic resonance point of view the
free photon criterion for localization is a very delicate one to achieve.
    This approach, however, overlooks an important aspect of the problem which
has far reaching consequences in the high density limit. In a sense it is the
fundamental theorem of solid state physics: Certain geometrical arrangements of

identical scatterers can give rise to large scale or macroscopic resonances. The most familiar example is the Bragg scattering of an electron in a perfectly periodic crystal. Such an effect is not given the required statistical weight by a disorder average which improperly averages over all positions of the scatterers. Consider for instance a fluctuating dielectric constant $\epsilon(x) - \epsilon_0 \equiv \epsilon_{fluct}(x) = \epsilon_1(x) + V(x)$ where $\epsilon_1(x) = \epsilon_1 \sum_G U_G \exp(iG \cdot x)$ is a perfectly periodic Bravais superlattice and $V(x)$ is a small perturbation arising from disorder. Here G runs over the appropriate reciprocal lattice and its value for the dominant Fourier component U_G is chosen so that the Bragg condition $k \cdot G = (1/2)G$ may be satisfied for a photon of wavevector k. Such a structure is attainable, albeit in a low dielectric contrast regime, with polyballs in suspension which exhibit fcc and bcc superlattice arrangements as well as a number of disordered phases. Setting $V(x) = 0$ for the time being, the effect of the periodic modulation or the photon spectrum may be estimated within a nearly–free–photon approximation. Unlike scalar electrons, there is a degeneracy between two possible photon helicity states, namely the right and left hand circularly polarized states, as well as the possibility of helicity flip scattering. For scattering by an angle θ, the amplitude for helicity flip scattering is given by $(1 - \cos\theta)/2$. It follows that when the photon wavevector k lies along the Bragg plane defined by the reciprocal lattice vector G, the allowed photon frequencies are given by

$$\frac{\omega}{c} = \frac{k}{\sqrt{\epsilon_0 \pm \epsilon_1 U_G}} \tag{9a}$$

and

$$\frac{\omega}{c} = \frac{k}{\sqrt{\epsilon_0 \pm \epsilon_1 U_G | 1 - \frac{1}{2}G^2/k^2 |}} \tag{9b}$$

The first pair of frequencies corresponds to the optical s–wave in which the polarization vector is perpendicular to the plane defined by the incident beam and its Bragg reflected partner. The second pair corresponds to the optical p–wave in which the polarization vector lies in the plane of scattering.

The existence or near existence of a gap in the photon density of states is of paramount importance in determining transport properties and especially localization. Such a possibility was completely overlooked in the derivation of the free–photon Ioffe–Regel condition which assumed an essentially free–photon density of states. In the vicinity of a band edge the character of propagating states is modified. To a good approximation the electric field amplitude of the wave is that of two counterpropagating free photons with a resulting amplitude modulation whose wavelength $\lambda_{envelope}$ diverges as the band edge is approached. Under these circumstances the wavelength which must enter the localization criterion is that of the envelope. In the presence of even very weak disorder, the criterion $2\pi l/\lambda_{envelope} \simeq 1$ is automatically satisfied as the photon frequency approaches the band edge frequency. It is useful as a first step to estimate the dielectric contrast required to produce an optical band gap in three dimensions. For concreteness consider an fcc superlattice. An estimate within the nearly–free–photon approximation follows from the condition that there exist frequencies ω which remain in the spectral gap defined by equations (9a) and (9b) as the wavevector k

is allowed to span the surface of the *fcc* Brillouin zone. Setting $U_G = 1$, it is apparent from these solutions that the lowest frequency state of the upper branch ($-$ sign) occurs at the center of the Bragg plane $k = G/2$ provided $\epsilon_1/\epsilon_0 \leq 1/3$ whereas for higher dielectric contrast ($\epsilon_1/\epsilon_0 > 1/3$) the corresponding minimum occurs along a circle defined by the intersection of the Bragg plane with the sphere $k^2 = G^2/(1 + \epsilon_0/\epsilon_1)$. This is a new feature arising from the vector nature of light. Although the analysis based on equations (9a) and (9b) ignores complications arising fron the intersection of two or more Bragg planes, it leads to the estimate that for $\epsilon_1/\epsilon_0 > .36$ a severe depression in the photon density of states (Fig.3) and perhaps even a spectral gap occurs near the frequency $\omega/c = (2\epsilon_1)^{1/2}G/(\epsilon_0 + \epsilon_1)$. The nature of light propagation just above this frequency is described by the associated photon dispersion relations. Defining q as a small deviation of the photon wavevector k from a point k_0 on the Bragg plane in the circular photonic valley $(k_0)^2 = G^2/(1 + \epsilon_0/\epsilon_1)$, it follows that to quadratic order the lowest lying photon branch has frequency

$$\omega^2/c^2 = E_c + A_1(q_1)^2 + A_2(q_2)^2; \quad (\epsilon_1/\epsilon_0 > 1/3). \tag{10}$$

Here, the components of q along the principal axes perpendicular to the circular valley are labeled q_1 and q_2, whereas the third component q_3 tangent to the circle describes degenerate states and does not enter the dispersion relation (10) at quadratic order. The coefficients are given by $E_c = 2(k_0)^2/(\epsilon_0 + \epsilon_1)$; $A_1 = (2/\epsilon_0)[((\epsilon_0)^2 + (\epsilon_1)^2)/((\epsilon_0)^2 - (\epsilon_1)^2)]$ and $A_2 = 2(3 - \epsilon_0/\epsilon_1)/(\epsilon_0 + \epsilon_1)$. The phase space available for photon propagation is accordingly restricted to a set of narrow symmetry related cones in k–space analogous to the pockets of electrons near a conduction band edge well known in semiconductor physics. The perturbative introduction of randomness $V(x)$ leads to a mixing of all nearly degenerate photon branches. This invariably causes a smearing in wavevector space of any sharp Bragg resonances as well as a filling in of the pseudogap in the photon DOS. It is highly plausible nevertheless that photons near the band edge frequency retain certain general features of the dispersion relation (10) and that coherent backscattering of light occurs by disorder–induced scattering within and between such valleys in phase space. The existence of the band gap guarantees the existence of strongly localized photonic band tail states at frequencies $\omega^2/c^2 <\sim E_c$, analogous to the Urbach band tail in electronic systems. If the disorder is a weak perturbation on the underlying band structure, the position of the associated mobility edge is given by $ql \simeq 1$ where q is defined in (10). The physical picture is entirely different from that described by the free–photon Ioffe–Regel condition in that it is now the wavelength $2\pi/q$ rather than the vacuum wavelength $2\pi c/\omega$ which enters the localization criterion.

The occurence of a photonic band gap guarantees the observability of localization. However, in real disordered systems it is likely that neither the band picture nor the single scattering approach provides a complete description. This is particularly evident in the fact that band gaps which may be produced for optical waves are at best very narrow and that in the presence of significant structural disorder what remains is merely a large depression or pseudogap in the DOS. In the limit of dilute uncorrelated scatterers, the microscopic resonance picture is entirely adequate whereas in the high density limit of highly correlated scatterers a macroscopic or band point of view is essential. Classical wave localization in three dimensions however occurs in an intermediate regime in which elements of both of

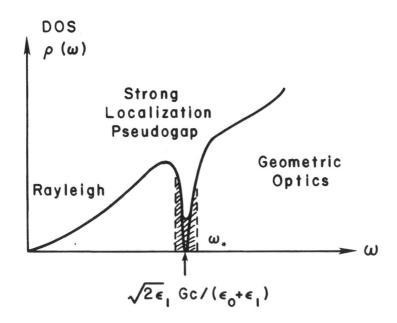

Fig.3   Photon density of states in a disordered superlattice exhibiting low
         frequency Rayleigh scattering and high frequency geometric optics
         extended states separated by a pseudogap of strongly localized photons.

these conceptually different approaches must enter but neither by itself is
complete. The two approaches are in this sense complementary.

4. OPTIMAL STRUCTURES FOR PHOTON LOCALIZATION: BEYOND
THE NEARLY FREE PHOTON APPROXIMATION

This qualitative picture of photon localization must be supported by quantitative
theoretical predictions for the location of gaps in the photon density of states.
Although the detailed bandstructure of a physical electromagnetic wave has yet to
be calculated, considerable insight may be gained from the example of scalar
waves. For scalar waves scattering from fcc, periodic array of dielectric spheres of
dielectric constant ϵ_a embedded in a uniform medium ϵ_b, it has been shown [13,14]
that for any fixed value of ϵ_a/ϵ_b there exists an optimum volume filling fraction f.
For $\epsilon_a/\epsilon_b >\sim 8$ a complete gap in the density of states occurs provided f is chosen
to be between 0.1–0.15. As f is varied from its optimum value, the size of the band
gap decreases rapidly. The existence of such an optimum is directly related to the
complementary mechanisms for localization, namely Bragg scattering and the Mie
resonance of individual spheres. When $f = f_{optimum}$, both of these mechanisms are
active at a single frequency whereas for $f \neq f_{optimum}$ the Bragg resonance and the
first Mie resonance occur at different frequencies.

A simple one dimensional example is instructive. Consider the scattering of a
classical scalar wave in $d = 1$ from a periodic array of square wells with refractive

index n_a, of radius a in a background $n_b = 1$ with lattice constant L. The volume filling fraction here is given by $f = 2a/L$. The analog of the Mie resonance is the condition of maximum wave reflection from a single square well. This occurs when precisely $1/4$ of wavelength fits into the well i.e., $\lambda/(4n_a) = 2a$. The Bragg condition is given by $2\pi/\lambda = \pi/L$. Combining these two conditions yields $f = 2a/L = 1/(2n_a)$. It may be verified by an exact bandstructure calculation that the condition $f = 1/(2n_a)$ does indeed give rise to the largest possible bandgap. Only in the very weak scattering limit $n_a \rightarrow 1$ does this condition reduce to that of maximum Fourier component of the scattering potential which occurs at $f = 0.5$. Precisely the same physics is operative in the scalar wave band structures in $d = 3$. By making use of tabulated Mie resonance for physical electromagnetic waves, this approach should yield an approximate "back of the envelope" estimate of the optimum scattering microstructre for a given dielectric contrast.

5. CONCLUSIONS

Photonic band gaps, pseudogaps and localization are of fundamental as well as applied interest. From a fundamental point of view electromagnetic propagation in lossless dielectrics provides the most direct experimental test of theories of transport in disordered media, unencumbered by finite temperature and manybody effects common to electronic transport. A detailed study of the critical point in three dimensions is in principle possible including a measurement of the scale dependence of the optical diffusion coefficient, time resolved pulse propagation and localization length. Also the accessibility of the critical point as a function of the static structure factor $S(k)$ of the disordered medium can be studied. In particular a precise reformulation of the Ioffe–Regel criterion for systems in which $S(k)$ is intermediate between being completely featureless to having Bragg peaks at specific values of k could be compared with experiment. A recent measurement by Drake and Genack [15] suggest that with greater control of structure and short range order, scale dependent optical diffusion in certain titania microstructures is very likely to be observed.

  Observable consequences of photon localization include enhanced backscattering and time delayed optical transmission through dielectric slabs. In addition to the macroscopic consequences are microscopic effects associated with radiation–molecule interactions. Inhibited spontaneous emission of light from atoms has been discussed by Yablonovitch [16] in the case of perfectly ordered dielectric microstructures. These considerations are likely to be important in disordered media possessing pseudo–gaps in the photon density of states. Kurizki and Genack [17] have also discussed the alteration of resonant energy transfer between molecules. Here the exchange of real as well as virtual photons is inhibited by the absence of electromagnetic modes at the relevant resonant energy. In the presence of a true photonic band gap, energy transfer is classically forbidden but must occur instead by means of photons tunnelling through the gap. In the presence of disorder, the dependence of energy transfer rates on interatomic separation may provide a direct measure of localization lengths. A detailed study of the quantum electrodynamics of atoms interacting with localized optical modes may also lead to fundamentally new phenomena in quantum optics.

REFERENCES

[1] S. John, Phys. Rev. Lett. 53 (1984) 2169; S. John, *ibid.* 58 (1987) 2486.
[2] P.W. Anderson, Philos. Mag. B52 (1985) 505.
[3] F. Urbach, Phys. Rev. 92 (1953) 1324; W. Martienssen, J. Phys. Chem. Solids 2 (1957) 257.
[4] S. John and M.J. Stephen, J. Phys. C17 (1984) L559.
[5] S. John, C. Soukoulis and M.H. Cohen and E. Economou, Phys. Rev. Lett. 57 (1986) 1777.
[6] B.I. Halperin and M. Lax, Phys. Rev. 148, (1966) 722.
[7] E. Abrahams and P.W. Anderson, D.C. Licciardello and T.V. Ramakrishnan, Phys. Rev. Lett. 42 (1979) 673.
[8] E.N. Economou, C.M. Soukoulis and A.D. Zdetsis, Phys. Rev. B30 (1984) 1686.
[9] S. John, H. Sompolinksy, M.J. Stephen, Phys. Rev. B27 (1983) 5592.
[10] See for instance M. Kerker, *The Scattering of Light and Other Electromagnetic Radiation* (Academic Press, New York, 1969).
[11] A.F. Ioffe and A.R. Regel, Prog. Semicond. 4 (1960) 237.
[12] S. John, Phys. Rev. B31 (1985) 304.
[13] S. John and R. Rangarajan, Phys. Rev. B38 (1988) 10101; R. Rangarajan, Bachelor's Thesis (Princeton University, 1988).
[14] K. Georgakis, Bachelor's Thesis (Princeton University, 1989).
[15] M. Drake and A.Z. Genack, Phys. Rev. Lett. 63 (1989) 259.
[16] E. Yablonovitch, Phys. Rev. Lett. 58 (1987) 2059.
[17] G. Kurizki and A.Z. Genack, Phys. Rev. Lett. 61 (1988) 2269.

Sajeev John was with Joseph Henry Laboratories of Physics, Princeton University, Princeton NJ; he is now with Department of Physics, University of Toronto, Toronto, Ontario, Canada M5S 1A7.

PHOTONIC BAND STRUCTURE

E. Yablonovitch

We employ the concepts of band theory to describe the behavior of electromagnetic waves in three dimensionally periodic face–centered–cubic (fcc) dielectric structures. This can produce a "photonic band gap" in which optical modes, spontaneous emission, and zero point fluctuations are all absent. In the course of a broad experimental survey, we have found that most fcc dielectric structures have "semi–metallic" band structure. Nevertheless, we have identified one particular dielectric "crystal" which actually has a "photonic band gap". This dielectric structure, consisting of 86% empty space, requires a refractive index contrast greater than 3 to 1, which happens to be readily obtainable in semiconductor materials.

1. INTRODUCTION

By analogy to electron waves in a crystal, light waves in a three dimensionally periodic dielectric structure should be described by band theory. Recently, the idea of photonic band structure [1,2] has been introduced. This means that the concepts of reciprocal space, Brillouin zones, dispersion relations, Bloch wave functions, van Hove singularities, etc., must now be applied to electromagnetic waves. If the depth of refractive index modulation is sufficient, then a "photonic band gap" could open up. This is an energy band in which optical modes, spontaneous emission, and zero point fluctuations are all absent.

It is interesting that the most natural real space structure for the optical medium is face–centered–cubic (fcc), which also happens to be the most famous atomic arrangement in crystals. The contrasts between electronic and photonic band structure are striking:

(i)   The underlying dispersion relation for electrons is parabolic, while that for photons is linear.

(ii)   The angular momentum of electrons is 1/2, but the scalar wave approximation is frequently made; in contrast, photons have spin 1 and the vector wave character plays a major role in the band structure.

(iii)   Band theory of electrons is only an approximation due to electron–electron repulsion, while photonic band theory is essentially exact since photon interactions are negligible.

The possible applications of such a "photonic band gap" are quite tantalizing. In addition to quantum electronic applications, such as spontaneous emission inhibition [1], there have also been proposals [2,3] to study mobility edges and Anderson localization within such a forbidden gap. Furthermore, Kurizki *et al.* [4] have shown that atomic and molecular physics is profoundly modified in a volume

W. van Haeringen and D. Lenstra (eds.), Analogies in Optics and Micro Electronics, 117–133.

of space in which "vacuum fluctuations" are absent. In particular, the interatomic potential of homonuclear diatomic molecules, as well as many other atomic physical properties are severely modified in such a spatial region.

Since we are only at the threshold of such research, we have elected to do our initial experimental work at microwave frequencies, where the periodic dielectric structures can be fabricated by conventional machine tools. Furthermore, this has enabled us to use sophisticated microwave homodyne detection techniques to measure the phase and amplitude of the electromagnetic Bloch wave functions propagating through the "photonic crystal".

Earlier work [1,2] had indicated that it was desirable for the Brillouin Zone in reciprocal space to be as near to spherical as possible. Among possible 3–dimensional periodic structures, this had suggested that face–centered–cubic (fcc) dielectric geometry would be optimal for achieving a "photonic band gap". The lowest order Brillouin Zone for the fcc structure happens to be nearer to spherical than the Brillouin Zone of any other common crystal structure. In the absence of any further theoretical guidance, we adopted an empirical, Edisonian approach. Literally, we used the cut–and–try method. Dozens of face–centered–cubic structures were painstakingly machined out of low–loss dielectric materials. These structures, which might be called "crystals", were roughly cube–shaped and contained up to ~ 8000 "atoms". In some cases the "atoms" were dielectric spheres, in other cases the "atoms" were spherical cavities filled with air ("spherical air–atoms"), the interstitial space consisting of dielectric material. The atomic volume filling fraction was varied from 11% to 86%. Refractive index contrast was varied between 1.6 to 1 and 3.5 to 1. The propagation of electromagnetic waves through these structures was then carefully investigated. This tedious cut–and–try approach was very time consuming, but it helped to ensure that no possibilities were overlooked.

The main conclusion of this paper is that a photonic band gap can indeed be achieved in 3–dimensional dielectric structures, but it requires an index contrast nearly 3.5 to 1. The early predictions had been much more optimistic, anticipating a gap opening up at index contrast 1.21 to 1 in one case [1] and 1.46 to 1 in the other [2]. In the experiments, all the test structures except one turned out to be "semi–metals" and only one particular geometry having index contrast 3.5 to 1 gave rise to a "semiconductor" with a true photonic band gap. In semi–metals, the valence band in one section of the Brillouin Zone has an energy overlap with the conduction band in a different section of the Brillouin Zone. A true bandgap, as in a semiconductor, requires a forbidden band of energies *irrespective of the propagation direction* in reciprocal space.

Fortunately, crystalline Silicon and other semiconductors, are excellent infrared optical materials, providing refractive indices ~3.5. Therefore the optimal structure we have found in the microwave experiments can be scaled down in size to provide a photonic band gap in the near infrared.

2. EXPERIMENT

The investigations employed the experimental arrangement shown in Fig.1. A monopole antenna (a 6 mm pin above a ground plane) launches a spherical wave down a long anechoic chamber built of microwave absorbing pads. The wave–front becomes approximately planar by the time it reaches the fcc dielectric structure at the opposite end of the chamber. (Henceforth the fcc dielectric structure will be called the "crystal"). Only the plane wave passing directly through the crystal can

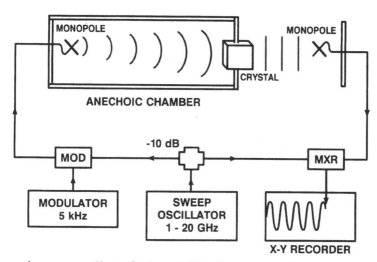

Fig.1   A sweep oscillator feeds a 10dB splitter. Part of the signal is modulated (MOD) and then propagated as a plane wave through a face–centered–cubic dielectric crystal. The other part of the signal is used as local oscillator for the mixer (MXR) to measure the amplitude change and phase shift in the crystal. Between the mixer and the X–Y–recorder is a lock–in amplifier (not shown)

1-D PERIOD STRUCTURE

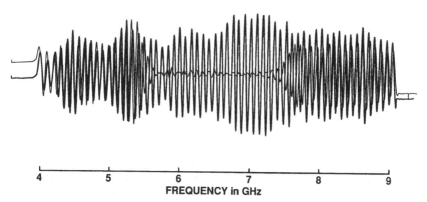

Fig.2   The interference pattern produced when a microwave signal being transmitted between antennas interferes with a local oscillator wave. The heavy line is a reference fringe pattern for the case of transmission through an empty space. The envelope fluctuations are due to variations in antenna efficiency. The lighter line is for transmission through 9 polystyrene plates spaced at one–quarter wavelength. A bandgap opens up between 5.5 GHz and 7.5 GHz. Furthermore, on either side of the bandgap there is ~ 180° phase shift.

be seen by the receiving monopole. A microwave oscillator feeds the homodyne detection system with a frequency sweep from 1 to 20 GHz. This home–built system resembles an optical Mach–Zender interferometer. (Alternatively, an HP–8510 Network Analyzer, was also used to perform the amplitude and phase measurements.) The X–Y recorder plots the interference pattern as a function of microwave frequency.

An example of such a series of interference fringes is shown in Fig.2 for the transmission of microwaves through a quarter–wavelength stack of 9 polystyrene plates (index 1.6, thickness 6 mm) spaced by quarter–wavelength air layers. In such a simple one–dimensional layered structure, a very pronounced stop–band or 1–dimensional bandgap occurs around 6.5 GHz. Two curves are present in Fig.2. The heavy curve is a reference interference fringe pattern showing the antenna transmission function when the stack of plates is removed. Variations in the wave envelope as a function of frequency are simply due to variations in the antenna efficiency. The lighter curve represents the experimental signal, the interference fringe pattern observed when the quarter wavelength stack of plates is inserted into the path of the microwave beam. Over a 2GHz band, centered on 6.5 GHz, the transmitted beam is severely attenuated. This is the stop–band.

Notice also the phase shift with respect to the reference beam in Fig.2. At 4.5 GHz and at 9 GHz, on either side of the stop–band, the signal fringes and the

Fig.3   A photograph of the 3–dimensional periodic structure which had a "photonic band gap". This face–centered–cubic crystal consisted of spherical air–atoms which were *larger* than close–packed. The resulting overlap of the air–atoms allows us to see visible light transmitted all the way through along the <111> direction. The spherical air–atoms occupy 86% of the volume. The interstices between the atoms are filled with dielectric material of refractive index 3.5. The overall structure is 86% air and only 14% solid material.

reference fringes are essentially in phase. The opposite effect occurs at the band edges. Due to group velocity dispersion, the signal beam is shifted roughly 180° compared to the reference beam at the upper and lower band–edge frequencies, 5.5 GHz and 7.5 GHz respectively. The relative phase shift is linearly proportional to wave–vector k. In principle, with an apparatus of this type, it is possible to map out the full dispersion relation, i.e. frequency versus wave–vector or ω vs k. Similarly, commercial equipment such as the HP–8510 Network Analyzer, can read out directly in terms of group velocity versus frequency.

Our most interesting crystal, exhibited in Fig.3, is the one which has a photonic band gap. Its structure is most unusual. The spherical atoms consist of air, while the space between the atoms is filled with a dielectric material. This commercial low–loss dielectric material, Emerson & Cumming Stycast–12, has a microwave refractive index = 3.5. The volume fraction occupied by the spherical air–atoms is 86%. In fcc close–packed structures, the atomic volume is only 74%. Therefore the atomic spheres in Fig.3 are actually "closer than close–packed", i.e. they overlap slightly. Due to the overlapping atoms, it is possible to see all the way through the crystal along certain directions. The bright spots of light emerging on the top surface of the crystal in Fig.3 are being channeled from below along the <111> direction.

Crystals consisting of spherical air–atoms are relatively easy to fabricate. A series of hemispheres are drilled on one face of a dielectric plate by a numerically controlled machine tool. On the opposite face of the plate an offset series of

Fig.4   A photograph of a 3–dimensional face–centered–cubic crystal consisting of Al_2O_3 spheres of refractive index 3.06. These dielectric spheres are supported in place by the blue foam material, refractive index 1.01. These spherical dielectric atom structures failed to show a "photonic band gap" at any volume fraction.

hemispheres are drilled. Then many of these plates are simply stacked up so that
the hemispheres face one another, forming spherical air–atoms. The volume
fraction is varied changing the hemisphere diameter.

The beauty of spherical air–atom crystals is that they are self–supporting. The
more obvious fcc structure consisting of dielectric spheres is self–supporting only
for the case of close–packing. For any smaller volume packing fraction, the
dielectric spheres must be supported in position. The dielectric spheres consisted of
polycrystalline Al_2O_3, 6 mm in diameter with a microwave refractive index = 3.06.
The volume fraction was varied by changing the sphere spacing. The dielectric
spheres were supported by thermal compression molded dielectric foam of
refractive index $\sim$ 1.01. Precision molds were built of Aluminum jig plates having 6
mm diameter steel ball bearings embedded in them. Dielectric foam pads were
molded at 95°C, with the molds released at 40°C. The hemispherical depressions in
the molded foam were then filled with the Al_2O_3 spheres and the structure built up
into many layers. Fig.4 is a photograph of such a crystal consisting of dielectric
spheres supported by the blue dielectric foam.

The philosophy behind our experiments is to map out the frequency versus
wave vector dispersion relations for a whole series of 3–dimensional fcc crystals.
For each crystal it becomes necessary to explore all the different angles in
reciprocal space. The interference fringe pattern in Fig. 5 is an example of such a
measurement on the 86% spherical air–atom crystal of Fig.3. These fringes are
produced in the homodyne detection system by an electromagnetic wave
propagating toward the L–U line of the hexagonal L–plane in reciprocal space.
This wave was predominantly s–polarized, i.e. it was polarized parallel to the
X–plane. Two important items emerge from Fig.5. The lower gap edge frequency
and the upper gap edge frequency. The lower edge was defined by the sudden drop
in microwave transmission relative to a reference scan with the crystal absent. The
upper edge is defined by the frequency at which the transmitted signal recovers.

These two frequencies define band edges, but these band edges do not
necessarily fall at exactly the same point on the surface of the Brillouin Zone. The
reciprocal space position of these frequency edges is determined by momentum
conservation between the external and internal electromagnetic waves. In our
experiments, the incoming plane wave was generally incident on the <100> face of
the crystal. (The <100> face is by definition perpendicular to the <100>
momentum direction.) Upon transmission through the crystal surface, only the
component of wave vector momentum parallel to the surface plane is conserved.
This is very similar to the principle which leads to Snel's Law. In our geometry it
means that component of wave vector momentum perpendicular to the <100>
direction, $<0,k_y,k_z>$ is conserved upon entering the crystal. The position of the
frequency edge in reciprocal space is the point on the Brillouin Zone surface having
those identical momentum components k_y and k_z. In our experiments, the external
angle of incidence is held fixed as the frequency is swept. Therefore the upper and
lower gap edge frequencies will have different k_y,k_z and different Brillouin Zone
positions.

Due to the limitations on the external wave vector $<0,k_y,k_z>$ which could be
attained in air, some parts of the internal Brillouin Zone sometimes had to be
accessed by transmission through a pair of giant microwave prisms on either side of
the crystal. The prisms, over 15 cm square, were made of polymethylmethacrylate
(microwave refractive index 1.6) to increase the available external wave vector.

Step by step, the angle of incidence is varied, and the frequency of the first and
second band edges is mapped out on the surface of the Brillouin Zone. Our

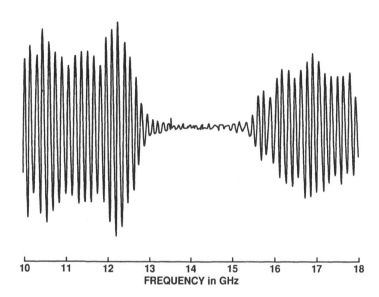

Fig.5   The forbidden gap observed on the crystal displayed in Fig.3 measured
         along the L–U line of the L–plane. The electromagnetic wave is
         polarized parallel to the X–plane (s–polarization).

measurements concentrated on the high symmetry planes X–U–L and X–W–K,
though lower symmetry points were also occasionally investigated. (See Fig.6 for a
description of the Brillouin Zone point labels). Unfortunately, with this method it
is not possible to learn much about the higher frequency bands above the two
lowest band edges. At higher frequencies, a superposition of allowed
electromagnetic modes can become excited by an incident plane wave. This makes
it very difficult to disentangle any higher band edges. Accordingly, we determine
only the first gap edge where transmission is cut off and the second where
transmission cuts on again. In keeping with the electron band structure analogy,
the first gap edge may be called the valence band and the second gap edge may be
called the conduction band.

The result of these measurements on our 86% spherical air–atom fcc dielectric
structure (photographed in Fig.3) is plotted in Fig.6. For electromagnetic waves,
two band structures must be shown, allowing for the two different polarization
states of electromagnetic waves. The forbidden bandgap in Fig.6 is filled in by
slanted lines. The lines that are slanted to the right fill in the bandgap for linear
polarization parallel to the X–plane, (mostly s–polarized). The left slanted lines
fill in the bandgap for the orthogonal linear polarization with a partial component
perpendicular to the X–plane, (mostly p–polarized). On the high symmetry planes,
X–U–L and X–W–K, the two linear polarizations are not expected to mix, and
therefore the linearly polarized antenna excites electromagnetic eigenstates of the
crystal. Off the high symmetry planes, the polarization eigenstates are no doubt
complex, and some type of elliptical polarization should be expected. No absolute
frequency units are given on Fig.6, since the frequencies should all scale with the

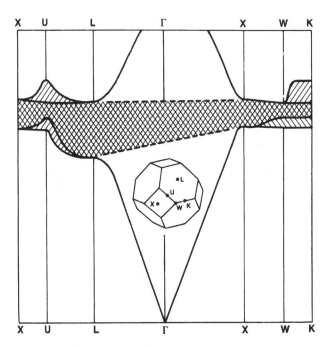

Fig.6   The experimentally observed photonic band structure in reciprocal space
         of the face–centered–cubic "spherical air–atoms" surrounded by the
         dielectric configuration of Fig.3. The right sloping lines represent
         polarization parallel to the X–plane, while the left sloping lines
         represent the orthogonal polarization which has a partial component out
         of the X–plane. The cross hatched region where both polarizations are
         forbidden in all directions in k–space is the "photonic band gap".

reciprocal linear size of the crystal spacing. On the actual crystal, the fcc unit cube
length "a" was equal to 12.7 mm and the forbidden gap was ∼ 1 GHz wide,
centered at 15 GHz.

   The band structure in Fig.6 plots frequency versus real wave vector. At
frequencies within the forbidden gap, the wave vector is pure imaginary, and it
measures the attenuation length within the crystal. Attenuation was generally very
strong within the bandgap, consistent with a 1/e attenuation length of only one or
two crystal unit cells. At points where the bandgap was very narrow, however,
particularly for p–polarized waves at the point U, the attenuation length was much
longer, ∼ 10 or 20 unit cells.

   All the other crystal structures that were fabricated and tested in our
experiments produced "semi–metals" rather than "photonic band gaps". Most
frequently, the conduction band at the point L in the Brillouin Zone generally
tended to overlap in energy with the valence band at the points W and U. An
example of such a band structure is shown in Fig.7 for the case of 50% volume
fraction spherical air–atoms embedded in polystyrene of refractive index 1.6. In our
experiments we found that it was essential to start out with large forbidden gaps
at the points X and L at centers of the square and hexagonal facets of the Brillouin

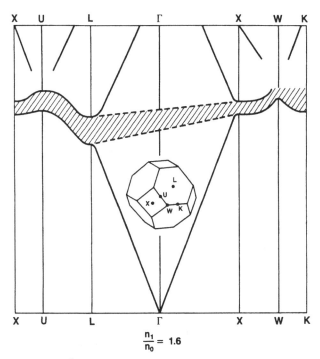

50% VOLUME FRACTION fcc AIR-SPHERES

$$\frac{n_1}{n_0} = 1.6$$

PREDOMINANTLY "P" POLARIZED

Fig.7   An example of a photonic band structure similar to that of a semi–metal. The fcc structure was 50% volume fraction air–atoms but was made of polystyrene, refractive index only 1.6. The conduction band at the L–point overlaps the valence band at the U and W points. In order to get a bandgap, both the volume fraction of atoms must be increased and the index contrast must be increased.

Zone. Invariably if bandgaps at the X and L points were inadequate, band overlap would become established at peripheral points between facets, such as U, W, or K. In our experimental survey, a semi–metallic band structure occurred in all cases but one.

Let us now analyze more of the band structure properties as a function of volume fraction and structure type. Fig.8 gives the index of refraction of the different crystal structures as a function of volume filling fraction. The effective refractive index is defined as $c/a\nu_X$, in terms of the center frequency, ν_X, of the X–point gap. We found that the X–point center frequency was a good extrapolation of the low frequency dispersion out to the X–point. Therefore $c/a\nu_X$ is a good surrogate representation of the long wavelength refractive index of the composite structure represented by these fcc crystals. For both the spherical

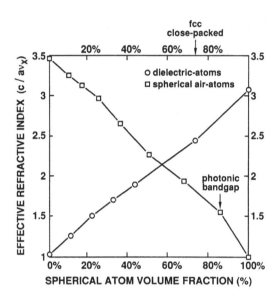

Fig.8   The effective long wavelength refractive index, $c/a\nu_X$, measured on two
         basic crystal structures of various volume fractions where c is the speed
         of light in vacuum, a is the length of the unit cube, and ν_X is the center
         frequency of the X–gap. Spherical dielectric atoms are represented by
         ($\square$)–points and spherical air–atoms are represented by ($\circ$)–points. Fcc
         close–packing occurs at 74% volume fraction. Of all these structures,
         only the one marked "photonic bandgap" had a forbidden frequency
         band at all directions in reciprocal space.

air–atoms and the dielectric sphere atoms, a simple linear interpolation of
refractive index with volume fraction seems to describe the experimental indices in
Fig. 8. Neither of the effective medium theories, not Maxwell–Garnett nor
Bruggeman were any more accurate in modelling Fig.8. Only the one structure
marked "photonic bandgap" on Fig.8 had a forbidden gap all the way around the
Brillouin Zone. All the others were semi–metals as mentioned earlier.
    In view of the importance of having large gap widths at the L–point and at the
X–point, we present those results in Fig.9. The measured gap width normalized to
ν_X is plotted against spherical air–atom volume fraction. All the results in Fig.9
were taken on Emerson & Cumming Stycast–12, a material with index contrast 3.5
to 1 relative to air. It is clear from Fig.9 why the 86% volume fraction spherical
air–atom structure has an overall photonic bandgap. Its gap widths are far larger
than any of the others. Notice the unusual behavior of the X–gap as a function of
spherical air–atom volume fraction. At around 68% volume fraction, the X–gap
width becomes undetectable while at the higher volume fraction of 86% it rises
again to a very large value. While our measurements can only assign a positive
number to the gap width, we have elected to plot this last data point as a negative
quantity. As shown in Ref. 2, the gap width on any plane in reciprocal space is
proportional to the corresponding Fourier component of the dielectric constant. We

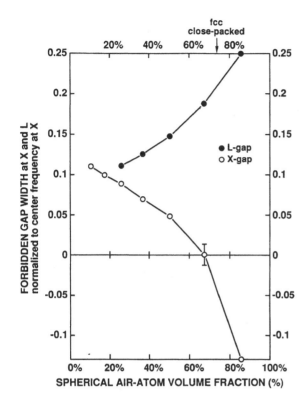

Fig.9   The forbidden gap width normalized to ν_X observed at the L–point ($\bullet$), and at the X–point (o) for a series of spherical air–atom fcc crystals of varying atomic volume fraction. The dielectric material between air–atoms had a refractive index of 3.5. We have elected to plot the 86% X–point gap width as a negative number since the <200> Fourier component has opposite sign on either side of the null point at 68% volume fraction.

believe that the zero gap at 68% volume fraction is a node at which the Fourier component of dielectric modulation changes sign.

For the X–plane Fourier component of the dielectric constant to change sign at a volume fraction around close packing is not entirely surprising. Fig.10 is a perspective view of the fcc close–packed structure. Remember that the spheres are air and the spaces between the spheres are filled with dielectric material. At the lower edge of the bandgap, the Bloch electric wave function tends to be concentrated in the high dielectric constant layers, while the upper edge electric field function tends to concentrate in the low dielectric constant layers. For the sake of definiteness let us concentrate on lower edge or valence band electric field function near the X–plane. The electric field will seek out those layers parallel to the cube faces in Fig.10 which have the highest dielectric constant. If the spherical air–atoms are much smaller than close–packed, the electric field will weave

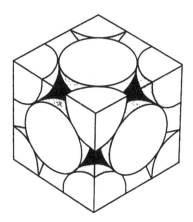

Fig.10   A close–packed face–centered–cubic structure. If the atoms are air–filled
           and smaller than close–packed, the electric field lines of the valence
           band edge X–gap mode will tend to weave between atomic layers. If the
           air–atoms are larger than close–packed, or if they are filled with
           dielectric material, the electric field lines of the valence band edge
           X–gap mode will tend to concentrate within the planes of atoms.

between the layers of atoms and avoid the atomic layers themselves. For spherical
air–atoms larger than close–packed, as can be seen from Figs.10 and 3, there is
little dielectric material left between the atomic layers and most of the remaining
material is in the cusp shaped volumes in the planes of atoms. In that case the
electric field tends to concentrate in the atomic layers themselves.

The plane containing the most dielectric material shifts from the inter–atomic
layers to the atomic layers as the spherical air–atom volume fraction goes up. As a
consequence, the amplitude of the Fourier component of the dielectric constant
goes through zero and changes sign. This explains the peculiar behavior of the
X–gap width in Fig.9, and justifies plotting the 86% data point as a negative
quantity. In that last case the valence band Bloch electric wave function tends to
concentrate in the planes of atoms along the cube faces.

By inspection of Figs.3 and 10, it is clear that there is no tendency for the
L–plane dielectric Fourier component to change sign. The electric field tends to
concentrate between the layers of atoms and there is no indication of a node in the
data. Therefore the L–gaps were all plotted as positive quantities.

The gap widths for spherical dielectric Al_3O_2 atoms are displayed in Fig.11.
The gap widths are all rather feeble, explaining why no overall photonic band gap
was observed for dielectric spheres. The L–plane gap was particularly weak. It
exhibited a polarization dependence even at the center of the hexagonal L–plane,
where s and p polarization are degenerate. The X–plane gap was stronger. There
were some indications that the 74% volume fraction X–gap data point should have
been plotted as a negative quantity as before. That data point was unusually
sensitive to tiny changes in packing geometry, possibly indicating that it was near
a node. In the absence of any additional information, we simply left the 74%
X–gap width as a positive number in Fig.11. The index contrast in this case was

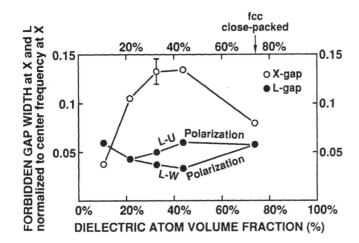

Fig.11   The forbidden gap width normalized to ν_X observed at the L–point ($\bullet$),
and at the X–point (o) for a series of fcc crystals made of Al_2O_3
spherical dielectric atoms, (refractive index 3.06). The atomic volume
fraction was varied by changing the size of the unit cube. Overall, the
L–point gap width was too feeble for the existence of a "photonic band
gap".

3.03 to 1, somewhat less than the spherical air–atom case. But that is not the
reason for the absence of a photonic band gap for the Al_3O_2 dielectric spheres.
Instead the absence must be attributed to the smallness of the L–plane gap at all
tested volume fractions.

3. THEORY

A starting point for the behavior of light waves in 3–dimensionally periodic
dielectric structures is derived from Maxwell's equations

$$-\nabla^2 E + \nabla(\nabla \cdot E) = \epsilon(x)\frac{\omega^2}{c^2} E, \tag{1}$$

where E is the optical electric field, c the speed of light, ω is the optical frequency,
which plays the role of an eigenvalue, and the geometry of the dielectric structure
is contained in the spatial dependence of dielectric constant $\epsilon(x)$. The dielectric
constant is related to the refractive index by $\epsilon \equiv n^2$. Eq. (1) resembles Schrödinger's
equation if $\epsilon(x)\omega^2/c^2$ is identified with the kinetic energy term $2m[E - V(x)]/\hbar^2$,
where m is the electron mass, $\hbar$ is Planck's constant divided by 2π, E is the total
energy eigenvalue and $V(x)$ is the potential energy.
    In optics we are generally restricting ourselves to positive dielectric constant
materials. Metals, which have negative dielectric constants, invariably have
significant dissipation as well, i.e. an imaginary component to $\epsilon(x)$. Among the
high quality optical materials, particularly semiconductors, a high positive

dielectric constant in the transparent region is accompanied by virtually no dissipation. This means that the kinetic energy term $\epsilon(x)\omega^2/c^2$ must always be positive in non–dissipative dielectric structures. If potential barriers are allowed, in which the kinetic energy is negative, it would be much easier to confine the wave functions, to produce localization, and to produce forbidden bandgaps. In optics with positive dielectric constant, it is very challenging to create a photonic band gap. For electrons, in a tight binding model for example, a forbidden bandgap occurs right from the outset. This is the main reason why only one dielectric crystal, out of the many we tested, had a photonic band gap.

As shown in Ref.2, the vector character of (1) also contributes to the difficulties of creating a photonic band gap. Waves which are p–polarized relative to the local Brillouin Zone surface interact more weakly. Vector wave equations are extremely difficult to solve [5]. This is a pity, since for solid state band theorists, our dielectric structures represent the classic muffin–tin–potential. In the scalar wave case, this famous potential would be soluble. The various methods for solving wave equations in 3–D periodic media are reviewed in Chapter 9–11 of Ashcroft & Mermin [6]. In view of the relatively small gap widths we have found experimentally, and the difficulty of adapting some of the more sophisticated band theory methods to vector waves, we will analyze our data in terms of the "nearly free photon model". This is analogous to the well–known "nearly free electron model", and is appropriate so long as the wave functions are weakly perturbed plane waves. The widest relative gap width, which measures the strength of perturbation, occurred at the L–point in the 86% spherical air–atom crystal and was only ~ 1/4. All other gap widths in all the crystals were at least a factor 2 smaller, lending reasonable credence to the "nearly free photon model".

The "nearly free photon model" predicts a splitting of bands at the Brillouin Zone surfaces

$$\omega_\pm = \omega_0 \left[1 \pm \frac{n_G}{n_0} \right], \tag{2}$$

where $\omega_\pm$ are the angular frequencies of the upper and lower band edges and ω_0 is the center frequency at that point on the Brillouin Zone surface, n_G is the Fourier component of the refractive index corresponding to the reciprocal lattice vector G defining that face of the Brillouin Zone and n_0 is the mean refractive index. For the case of a layered structure with refractive index contrast Δn between layers, the corresponding $n_G = \Delta n/\pi$. In three–dimensional structures, the refractive index contrast is not organized into simple layers, and the plane wave Fourier components n_G tend to be much weaker than $1/\pi$ times the index contrast Δn.

Eq.(2), as written, applies only to s–polarized electromagnetic waves with respect to the Bragg reflection planes on the Brillouin Zone surface. For p–polarized waves, the scattering efficiency is diminished by the projection of the polarization vector onto the new scattered direction. Accordingly, n_G is diminished by $\cos(2\theta) = 2\cos^2\theta - 1$, where θ is the angle of incidence onto the Bragg plane forming the Brillouin Zone surface. The largest angle in the fcc structure is $\theta = 39°$, subtended by the points L and W. In Yeh [7] the forbidden gap width is increased by a further angle dependent factor, $1/\cos^2\theta$. The reason for this additional factor is that they are considering the stop–band for a fixed angle of incidence. As explained earlier in the experimental section of this paper, the component of photon momentum parallel to the entry plane is conserved. Therefore the upper and lower band edge frequencies observed experimentally, appear at *different* points along the Brillouin Zone surface. The apparent gap width becomes larger by $1/\cos^2\theta$. By contrast, (2) gives the band edge frequencies and gap width at a *single* point on the Brillouin Zone surface.

The quantity n_G $/n_0$ plays the role of a dimensionless pseudo–potential in the "nearly free photon model" and we will give it the label V_G. For example, the pseudopotential corresponding to the $G = <111>$ on the hexagonal L–plane of the Brillouin Zone can be defined $V_1 \equiv n(<111>)/n_0$. The dimensionless pseudo–potential corresponding to the $G = <200>$ on the square X–plane of the Brillouin Zone can be defined $V_2 \equiv n(<200>)/n_0$. In general these pseudo–potentials can also be expected to depend on polarization angles as explained above.

The "nearly free photon model" becomes particularly interesting at points of degeneracy on the Brillouin Zone surface. For example, at the point W, four different plane waves are degenerate and they mix to produce eigenmodes which break the degeneracy. Similarly at the point U, three different plane waves mix to produce a superposition of plane waves. As shown in Ref.6, the solution requires the diagonalization of the following matrix

$$
\begin{bmatrix}
-\Delta\omega/\omega_W & V_1 & V_1 & V_2 \\
V_1 & -\Delta\omega/\omega_W & V_2 & V_1 \\
V_1 & V_2 & -\Delta\omega/\omega_W & V_1 \\
V_2 & V_1 & V_1 & -\Delta\omega/\omega_W
\end{bmatrix} = 0,
\tag{3}
$$

where ω_W is the center frequency at point W in the Brillouin Zone and $\Delta\omega \equiv (\omega - \omega_W)$ where ω are the new eigenfrequencies. The solutions to (3) are as follows

$$\Delta\omega/\omega_W = -V_2 \quad \text{(twice)}, \tag{4a}$$

$$\Delta\omega/\omega_W = V_2 \pm 2V_1. \tag{4b}$$

Of these four solutions, the two in (4a) are degenerate. This degeneracy is no accident, nor is it specific to the "nearly free photon model". It is quite general and arises directly from the group–theoretical properties of the W–point in the fcc structure. In the character tables given by Callaway [8], the W_3 representation is doubly degenerate. If solution (4a) were to be the lowest lying frequency at the point W, then a photonic band gap would be impossible, since valence and conduction bands would touch. This can be prevented in two ways: (i) V_1 can be larger than V_2. (ii) As shown in Fig.9, V_2 can be negative. Either case will ensure that at least one of the two solutions (4b) is below solution (4a).

The difference between the wave functions for solutions (4a) and (4b) can be visualized from Fig.10. The mixing of the plane waves along the cubic X–axis results in standing waves which peak either within the plane of atoms on the cubic face or between the planes of atoms. Depending upon whether the atoms are made up of air or dielectric spheres, one of these standing waves will be bonding (lower in frequency) and the other will be anti–bonding (higher in frequency). The $<200>$ standing wave peaking on the cubic face will, at the W–point, interact strongly with other standing waves in the $<111>$ directions, leading to solution (4b). For the other $<200>$ standing wave, having nodes on the fcc cube faces, the interaction with $<111>$ plane waves cancels out, leading to (4a).

Similarly, the matrix equation for the point U is

$$\begin{bmatrix} -\Delta\omega/\omega_U & V_1 & V_2 \\ V_1 & -\Delta\omega/\omega_W & V_1 \\ V_2 & V_1 & -\Delta\omega/\omega_U \end{bmatrix} = 0, \tag{5}$$

where ω_U is the center frequency at point U in the Brillouin Zone and $\Delta\omega \equiv (\omega - \omega_U)$. The solutions to (5) are as follows

$$\Delta\omega/\omega_U = -V_2/2, \tag{6a}$$

$$\Delta\omega/\omega_U = V_2/2 \pm [(V_2/2)^2 + 2(V_1)^2]^{1/2}. \tag{6b}$$

For solution (6a), as before, the effect of the <111> pseudo–potential cancels out.

We are now in a position to test (4) and (6) against the observed band structure at the points W and U in Fig.6. It is necessary to take polarization into account. Fortunately, as discussed earlier, in the high symmetry planes X–U–L and X–W–K, there exist two orthogonal linear polarization which are not expected to mix. At the U–point these divide neatly into s–polarized parallel to the X–plane and p–polarized parallel to the X–U–L plane. At the W–point the two polarizations are not purely s or p, but they are orthogonal nevertheless. One polarization is parallel to the X–W–K plane and the other is parallel to the X–plane. Each orthogonal polarization leads to its own matrix equation, (3) or (5), with its own values of V_1 and V_2 which must be adjusted by polarization angle dependent factors. For p–polarization at the U–point, V_2 is multiplied by $\cos(2\theta) = 7/9$ and V_1 is multiplied by $\cos(2\theta) = 1/3$. Substituting into (6), the calculations are in reasonable agreement with experiment for both polarization components of the measured band structure at the point U in Fig.6.

A similar calculation (requiring more trigonometry) at the W point leads to gap widths that are significantly larger than observed in Fig.6. We do not know the reason. Evidently, we may possibly have reached the limits of the "nearly free photon model" at the W point.

4. CONCLUSIONS

We have found that the idea of a "photonic band gap" can be experimentally realized in 3–dimensionally modulated dielectric structures. It required a refractive index contrast approaching 3.5 to 1 and the optimal structure consisted of 86% spherical air–atoms and only 14% dielectric material. All the other structures that we tested were "semi-metals" with the conduction band minimum at the L–point overlapping with the valence band maximum at the W or U point. In particular, this included all the tested structures consisting of spherical dielectric atoms.

The main reason for the difficulty in producing a photonic band gap was that the Fourier components of the index modulation in a 3–dimensional structure tend to be much weaker than in a simple 1–dimensional layered structure. In the spherical dielectric atom case the <111> Fourier components tended to be particularly weak. It should be possible to scale the optimal structure from microwave wavelengths down to the near–infrared by microfabrication in semiconductors like Silicon and GaAs which do possess the required refractive index.

We would like to thank Prof. S. John for numerous discussions and Prof. Gene Mele for advice regarding the group theory.

REFERENCES

[1] E. Yablonovitch, Phys. Rev. Lett. 58 (1987) 2059.
[2] S. John, Phys. Rev. Lett. 58 (1987) 2486.
[3] S. John, Comm. Cond. Matt. Phys. 14 (1988) 193.
[4] G. Kurizki and A.Z. Genack, Phys. Rev. Lett. 61 (1988) 2269.
[5] S. John and R. Rangarajan, Phys. Rev. B38 (1988) 10101.
[6] N.W. Ashcroft and N.D. Mermin, *Solid State Physics* (W.B. Saunders, Philadelphia, 1976).
[7] P. Yeh, *Optical Waves in Layered Media* (Wiley, N.Y., 1988) 36.
[8] J. Callaway, *Quantum Theory of the Solid State*, Part A (Academic Press, New York, 1974) Appendix C.

E. Yablonovitch is with Bell Communications Research, Navesink Research Center, Red Bank, N.J. 07701–7040, USA.

OPTICAL LEVEL CROSSINGS

J.P. Woerdman and R.J.C. Spreeuw

We discuss the behavior of two coupled optical modes, specializing to the case of a clockwise and a counter–clockwise wave in an optical ring with backscattering. Generally the coupling constant is complex–valued; the limiting cases of conservative and dissipative coupling are emphasized. When lifting the polarization degeneracy we deal with four optical modes, coupled by backscattering and birefringence. We also consider an active i.e. oscillating optical ring: this corresponds to adding a nonlinear mode–coupling and leads to a description of frequency–locking in a ring laser gyro. Finally we address the analogy between optical and electronic mode coupling in ring–type and lattice–type configurations ("band structures"). Here we emphasize dynamic effects and discuss the merits and limitations of optical modeling of phenomena such as Bloch oscillations and Zener tunneling.

1. INTRODUCTION

Our topic is the coupling of two modes of the electromagnetic field. Because of the one–to–one correspondence of a mode of the electromagnetic field with a harmonic oscillator we deal basically with the coupling of two harmonic oscillators. It may seem that so much has been said about this topic that nothing new remains to be added. This is largely true and in that sense our contribution is meant to be a didactic one. It reviews in particular the coupling of light waves traveling clockwise (*cw*) and counterclockwise (*ccw*) along an optical ring; the coupling is due to backscattering by a static dielectric perturbation (R) somewhere along the ring (Fig.1a). This configuration is basic for studies of frequency locking in laser gyros [1–3], photon band structure in ring resonators [4,5] and statistical properties of bidirectional ring lasers [6–8]. Usually, papers presented under these headings use different theoretical approaches. We intend to construct here a simple unifying framework indicating the connections.

Let us return to the optical ring of Fig.1a. We will consider only one transverse mode. The eigenfrequencies of the uncoupled *cw* and *ccw* waves are degenerate. This degeneracy can be lifted by varying a dynamic variable S which acts in a nonreciprocal way on the *cw* and *ccw* waves [4,5]. For this we may mechanically rotate the ring at angular frequency Ω, using the Sagnac effect ($S=\Omega$) [4]. We may also simulate rotation by applying a magnetic field B, using the Faraday effect ($S=B$) [5]. Crossing of the uncoupled eigenfrequencies as a function of S is shown in Fig.1b; it represents an optical level crossing.

The ring represents a macroscopic, topological periodicity since the circulating light encounters periodically the disturbance R; the periodicity length is the circumference of the ring, e.g. several meters. As shown in Fig.1c, two traveling light waves may also be coupled in an analogous way by a medium with a

135

W. van Haeringen and D. Lenstra (eds.), Analogies in Optics and Micro Electronics, 135–150.

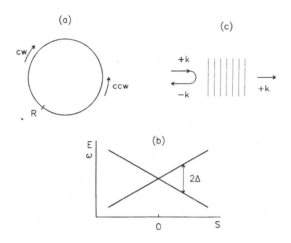

Fig.1   a) Optical ring resonator; *cw* and *ccw* waves are coupled by a
         backscattering element R. b) Crossing of eigenfrequencies as a function
         of the dynamic variable *S*. c) Distributed feedback structure couples
         waves $+k$ and $-k$.

microscopic, spatial periodicity [9,10]. An example of such a medium is a
multilayer dielectric mirror ("distributed feedback", DFB) the periodicity length
being on the order of a wavelength (i.e. $\approx 10^{-6}$ m). In the microscopic as well the
macroscopic case light waves are coupled if their wavelength satisfies a Bragg
condition. Our discussion will be mainly for the macroscopic case (i.e. a ring) since
in that case experimental implementation of the issues discussed below is much
easier.

    Because the present volume is on analogies, it is proper to stress the
(mathematical) analogy of an optical level crossing with other linear coupled–wave
problems in physics, chemistry or engineering. An instructive but by no means
complete list is as follows: coupled mechanical oscillators such as pendulums,
strings or organ pipes [11,12]; coupled electrical circuits [11]; flux–periodic
phenomena in small normal–metal rings [4]; surface plasmons on a grating [13];
elementary particle resonances [14,15]; optical Stark effect and Rabi oscillations
[16]; microwave and fiber–optic directional couplers [9]; atomic and molecular level
crossings [17,18]; ionization of highly–excited atoms [19]; chemical reaction
dynamics [20]. In some of these examples only static properties of the level crossing
are involved (i.e. level crossing spectroscopy) whereas in others emphasis is on
dynamic properties (i.e. diabatic versus adiabatic transitions). The reader may
map the framework discussed below on his own favorite 2×2 eigenvalue problem; in
the text which follows we will occasionally hint at such connections.

2. COUPLED–MODE THEORY

The equations of motion of a pair of linearly coupled harmonic oscillators can be
expressed in the canonical form [12]

$$dA/dt = -iHA \tag{1}$$

with A the column vector (a_1, a_2) which represents the oscillator or mode amplitudes a_i and H the dynamical matrix,

$$H = \begin{bmatrix} \Delta & W_{12} \\ W_{21} & -\Delta \end{bmatrix}. \tag{2}$$

Note that (1) is analogous to the Schrödinger equation. The diagonal elements give the (real) detunings $\pm\Delta$ of the unperturbed oscillators or modes (Fig.1b); we will assume a linear relation $\Delta \propto S$. The complex off–diagonal elements W_{12} and W_{21} represent the coupling rates. Diagonalizing H yields two normal modes with eigenfrequencies

$$\omega_\pm = \pm\sqrt{\Delta^2 + W_{12}W_{21}}. \tag{3}$$

It is useful to distinguish the case that the coupling produced by H is Hermitian (i.e. $W_{12} = W_{21}{}^*$) and the case that it is anti–Hermitian (i.e. $W_{12} = -W_{21}{}^*$). In the Hermitian case the normal–mode frequencies are

$$\omega_\pm = \pm\sqrt{\Delta^2 + W^2} \tag{4}$$

where $W = |W_{12}|$ (see Fig.2a). This represents so–called conservative coupling [1]; the coupling conserves $|a_1|^2 + |a_2|^2$, i.e. the total energy of the two modes (equivalent labels are "reactive" or "dispersive"). The eigenfrequencies are pushed apart by the coupling, opening up a forbidden frequency gap. In analogy with electronic band structure of semiconductors we will label the valley hyperbola in Fig.2a as "conduction band" and the mountain hyperbola as "valence band".

In the conservative case the normal mode wave functions can be expressed as

$$|\Psi_+\rangle = \cos\Theta\,|cw\rangle + \sin\Theta\,|ccw\rangle$$
$$|\Psi_-\rangle = \cos\Theta\,|ccw\rangle - \sin\Theta\,|cw\rangle \tag{5}$$

where the notation refers to the case of the optical ring. The "mixing" angle Θ is given by $\tan 2\Theta = W/\Delta$. As an example, for the case of a ring rotating at frequency Ω the Sagnac effect yields $\Delta = m\Omega$, where m is the mode index, i.e. the number of optical wavelengths along the ring [4,5]. For $\Delta = 0$ we have $\Theta = 45^0$ and therefore complete mixing. Thus at the extrema of valence and conduction bands the normal modes are standing waves which have a mutual phase difference of 90^0. The frequency splitting of these normal modes ($2W$) can be interpreted as a difference in dielectric polarization energy. Consider for instance a glass fiber ring with a thin air gap acting as the dielectric perturbation R; in that case the valence band standing wave has a node and the conduction band standing wave an anti–node in the air gap (Similarly, in a semiconductor the electronic band gap can be interpreted as a difference in Coulomb energy of two mutually dephased (90^0) standing electronic waves in an ionic lattice). When the ring is set into rotation

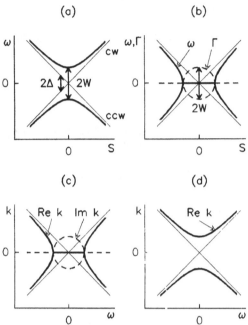

Fig.2   a) Normal–mode frequencies of two conservatively coupled modes in
optical ring resonator. b) Normal–mode frequencies and dampings of
two dissipatively coupled modes in optical ring resonator. c) Dispersion
and damping of two modes which are conservatively coupled by
distributed feedback ("index coupling"). d) Dispersion of two modes
which are dissipatively coupled by distributed feedback ("gain
coupling").

(Δ or $\Omega = 0 \rightarrow \infty$) the normal mode wave functions gradually transform into the
unperturbed traveling wave modes $|cw>$ and $|ccw>$.

In the anti–Hermitian case the coupling is called dissipative [1] (or,
alternatively, "resistive" or "absorptive") and the normal–mode frequencies are

$$\omega_\pm = \pm \sqrt{\Delta^2 - W^2}. \tag{6}$$

The eigenfrequencies are now pulled together by the coupling (Fig. 2b). For
detunings smaller than the dissipative coupling, i.e. $|\Delta| < W$, $\omega_\pm$ is purely
imaginary so that we deal with two normal modes with the same frequency
("frequency locking") but different dampings; for $\Delta = 0$ the dampings are,
respectively, $+W$ and $-W$ (Fig.2b). A damping $-W$ corresponds to gain; however,
this artefact disappears if damping of the unpertubed modes $|cw>$ and $|ccw>$ is
taken into account (see Sec.3). At the bifurcation points $\Delta = \pm W$ locking is broken
and frequency splitting starts; this bifurcation condition may be compared with the
critical–damping condition of a single harmonic oscillator, $\omega_0 = \gamma$.

In the dissipative case the normal–mode wave functions are standing waves for $|\Delta| < W$, their phase difference decreasing from 90^0 at $\Delta = 0$ to 0^0 at $|\Delta| = W$. For $|\Delta| > W$ traveling wave character is admixed and the normal modes transform again asymptotically into the unperturbed eigenmodes $|cw>$ and $|ccw>$. In the dissipative case the normal modes are *nonorthogonal* except for $|\Delta| = 0$ and $|\Delta| = \infty$; in the conservative case the normal modes are orthogonal for all Δ.

The present discussion is easily generalized to mode couplings in between conservative and dissipative; examples of $\omega(S)$ curves for this intermediate regime are given in Ref. [12]. It follows directly from (3) that only purely dissipative coupling leads to degeneracy of the normal–mode frequencies for a finite range of S–values (Fig.2b).

It is interesting to compare the present treatment with that of one–dimensional coupling of a left– and a right–traveling wave, with wave vectors $+k$ and $-k$, by a spatial grating (Fig.2c) [10]. Starting point is then the spatial analogue of (1), i.e. t should be replaced by z; solving again the eigenvalue problem leads to a dispersion diagram $k(\omega)$ (instead of $\omega(S)$). Conservative coupling corresponds to an index grating and dissipative coupling to a gain (or absorption) grating. An important difference with the case of the optical ring is that the coupling represented by the dynamical matrix H is anti–Hermitian for the conservative case and Hermitian for the dissipative case. For example, in the case of conservative coupling $|a_1|^2 - |a_2|^2$ is conserved, i.e. there is no spatial energy source or sink in the DFB structure. This difference is due to the fact that in the DFB case the coordinates of the uncoupled modes progress in opposite directions (z and $-z$), contrary to the case of the optical ring, where the uncoupled modes have a common coordinate t. As a consequence, the shape of $k(\omega)$ is given by Fig.2c for conservative (index) coupling and by Fig.2d for dissipative (gain) coupling [10].

In the zoo of level crossings mentioned at the end of Sec.1 the conservative and dissipative variety are both present. Level crossings in microscopic, i.e. (sub)atomic systems are usually of the purely conservative type. One notable exception is the (partly) dissipative coupling of a K^0 meson to its antiparticle $\overline{K^0}$ through interaction of their pionic decay products $\pi^+\pi^-$; the corresponding normal modes are a long–lived K_L^0 and a short–lived K_S^0 resonance [15], corresponding to $\gamma_- = 0$ and $\gamma_+ = 2W$ (see Sec.3).

3. INCLUSION OF DAMPING

We now introduce damping rates γ_i of the unperturbed modes. Let us assume first $\gamma_1 = \gamma_2 = \gamma$ so that

$$H = \begin{bmatrix} \Delta - i\gamma & W_{12} \\ W_{21} & -\Delta - i\gamma \end{bmatrix}. \tag{7}$$

This corresponds to adding a term $-\gamma A$ to the right–hand side of (1). Solving the eigenvalue problem yields $\omega_\pm = -i\gamma \pm (\Delta^2 + W_{12}W_{21})^{1/2}$, i.e. the same $\omega(S)$ curves as before, but each normal mode acquires a damping γ. For the case of dissipative coupling with strength W we have to assume that the dampings of the individual oscillators are $\gamma_1 = \gamma_2 \geq W$ since otherwise, as mentioned in Sec.2, one of the normal modes would have gain. In practical examples of dissipative coupling the coupling element involved also damps the oscillators separately so that $\gamma_1 = \gamma_2 = W$; as a result the normal–mode dampings (at $\Delta = 0$) are $\gamma_- = 0$ and $\gamma_+ = 2W$.

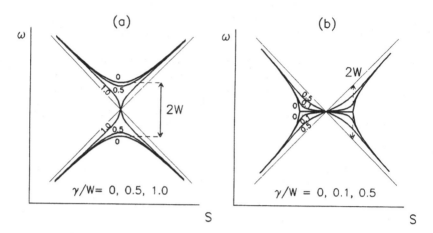

Fig.3   a) Normal–mode frequencies of two conservatively coupled modes, one
         of which is damped $(\gamma_1 = 2\gamma > 0,\ \gamma_2 = 0)$. Results are given for $\gamma/W =$
         0, 0.5 and 1.0. b) Idem for dissipative coupling, $\gamma/W = 0$, 0.1 and 0.5.

An interesting situation arises if the damping is asymmetric, i.e. $\gamma_1 \neq \gamma_2$.
Consider first an extreme example, $\gamma_1 = 2\gamma$ and $\gamma_2 = 0$. This example has been
known for a long time in the context of atomic level crossings [21]. Solving the
eigenvalue problem yields the same $\omega(S)$ curves as before if $\gamma << W$. For
increasing γ the $\omega(S)$ curves are deformed in a cusp–like fashion (Fig.3). In the
conservative case the frequency splitting vanishes for $\gamma > W$; in the dissipative
case the frequency degeneracy vanishes as soon as $\gamma > 0$. The general case $\gamma_1 \neq \gamma_2$
leads to identical curves, $|\gamma_1-\gamma_2|$ now playing the role of γ [18].
    A natural example of cusping occurs in the "dressed–state" description of the
optical Stark effect i.e. a description in terms of a conservative coupling of an
atomic transition and a near–resonant electromagnetic field [16]. In this case the
atomic transition is usually damped (due to spontaneous emission) whereas the
field mode is not.
    Note that in the case of two optical modes coupled by a DFB structure the
situation is reversed again [10]; in that case equal damping or gain leads to
maximum cusping whereas cusping is absent if the modes have opposite damping.

4. REALIZATION OF COMPLEX 2–MODE COUPLING IN A RING

For an optical ring resonator with partially backreflecting elements inside or
outside the resonator (Fig.4) the coupling rate W_{12} is related to the amplitude
reflection coefficient R_{12} by $iW_{12} = R_{12}\,c/L$, where R_{12} is generally a complex
number and L is the circumference of the ring. Here we assume $|R_{12}| << 1$. Note
that conservative coupling corresponds to $R_{12} = -R_{21}{}^*$ and dissipative coupling to
$R_{12} = R_{21}{}^*$ [7]. We will briefly discuss the (complex) coupling obtained in the
configurations shown in Fig.4, quoting results only since the actual calculations are
straightforward.
    First we consider a coupling element $inside$ the ring resonator (Fig.4a). A
Fabry–Perot etalon E, e.g. a solid glass plate of optical thickness nd, aligned

perpendicular to the mode axis, yields conservative coupling [22]; an air gap in a glass fiber ring has the same effect [5]. The strength W of this conservative coupling depends on the interference of the Fresnel reflection amplitudes at the two glass–air interfaces, $R_{12}^{(tot)} = R_{12}^{(1)} + R_{12}^{(2)}$. For a resonant etalon, i.e. $2nd = N\lambda$, this interference is destructive, i.e. $R_{12}^{(tot)} = 0$, due to the π phase–shift upon reflection from an optically dense medium. For an anti–resonant etalon, i.e. $2nd = (N + \frac{1}{2})\lambda$, interference is constructive; in this case $R_{12}^{(tot)} = (n^2-1)(n^2+1)$, which equals 0.38 for a glass etalon ($n = 1.5$), The strength of the (conservative) coupling is thus easily varied by changing nd (or λ). If one of the etalon faces (say 2) is anti–reflection coated interference does not occur and we simply have $R_{12}^{(tot)} = R_{12}^{(1)}$.

Dissipative coupling by intra–ring elements may be realized by having a localized loss somewhere along the ring. It may also be realized by having effective backreflection of mode 1 into mode 2 by scattering from small lossless dielectric particles (diameter a), e.g. dust particles on the resonator mirrors [1,2]. Coupling due to scattering from particles is purely dissipative in the Mie limit, $a >> \lambda$; in the Rayleigh limit, $a << \lambda$, the coupling is again conservative due to the inherent angular isotropy of Rayleigh scattering [1]. Constructive c.q. destructive interference of backscattering amplitudes may also occur in the dissipative case, for instance if we deal with dust particles located at different resonator mirrors.

Now we turn to coupling by extra–cavity elements [6,7]. Two mirrors are required in this case to realize R_{12} and R_{21}; they may be used in the configuration of Fig.4b or in that of Fig.4c. An advantage of extra–cavity coupling in general is that the conservative and dissipative cases may be realized using the same optical elements. For instance, in the configuration of Fig.4b the coupling changes periodically from conservative to dissipative and back, as a function of L, with a period $\lambda/2$, the strength W of the coupling remaining constant. In this case the mode coupling implies the appearance of two new traveling waves leaking to the outside world (see downward pointing arrows in Fig.4b). The phase relation of these waves is associated with the nature of the coupling (conservative or dissipative).

The mirrors M_1 and M_2 may also form a folded Fabry–Perot interferometer (Fig.4c). If this Fabry–Perot is loss–free the coupling is purely conservative, its strength being a periodic function of L. Resonance corresponds to maximum coupling (i.e. $|R_{12}| = |R_{21}| = 1$) and anti–resonance to minimum coupling. If the folded Fabry–Perot is lossy, e.g. if the mirrors M_1 and M_2 are partially transmitting, the coupling changes periodically as a function of L, from conservative (anti–resonance) to dissipative (resonance) and back, the period being $\lambda/2$; in this case the strength W of the coupling displays the Fabry–Perot resonances.

Since the normal–mode spectrum depends on $W_{12}W_{21}$, it depends only on the distance (L) between mirrors M_1 and M_2 and not on their individual positions. Similarly, in the case of intra–cavity coupling the normal–mode spectrum is independent of the position of etalon E. The positions of M_1 and M_2 (and of E) do however affect the phase of W_{12}. For the case of conservative coupling this leads, by analogy with electrical and mechanical oscillators, to a distinction between "inductive" and "capacitive" coupling [11], or "springy" and "massy" coupling [12]. For the two standing–wave normal modes in the conservatively coupled optical ring resonator the phase relations change periodically, as a function of position along the ring, from capacitive to inductive and back, the period being $\lambda/2$.

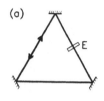 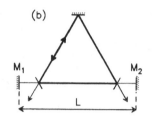 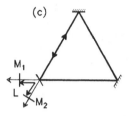

Fig.4   a) Intracavity etalon E couples cw and ccw waves. b,c) Extracavity
mirrors M_1 and M_2 couple cw and ccw waves.

5. LIFTING THE POLARIZATION DEGENERACY

So far we have assumed polarization degeneracy of the cw and ccw modes in the
optical ring, i.e. we have neglected birefringence of the ring elements[1]. If we do add
birefringence, linear and/or circular, the polarization degeneracy is lifted and we
have a problem of 4 modes, namely cw, ccw, E_x and E_y, interacting via 2
couplings, namely backscattering and birefringence. Instead of 2×2 we deal now
with 4×4 matrices. The birefringent optical ring resonator represents in fact an
optical implementation of wave mechanics in a 4–dimensional state space[2]. The
theoretical description proceeds as follows [23]. Consider a generalized ring
resonator (Fig.5a) composed of backscattering elements (isotropic or anisotropic),
birefringent elements (linear or circular), free propagators (isotropic or
anisotropic), Farady rotators, etc. Each element is represented by a transmission
matrix M_1 which relates the two complex Jones vector fields at its left to those at
its right (Fig.5b),

$$\begin{bmatrix} C \\ D \end{bmatrix} = M_i \begin{bmatrix} A \\ B \end{bmatrix}, \tag{8}$$

i.e. M_i is a 4×4 matrix with complex elements, some of which may depend on the
dynamic variable S. Introducing the transmission matrix of one roundtrip, $M \equiv
\Pi_i M_i$, and solving the eigenvalue equation $\det(M-1) = 0$ yields the 4 normal–mode
frequencies $\omega_i = \omega_i(S)$[3]. A few $\omega_i = \omega_i(S)$ diagrams for some particular optical
rings are shown in Fig.6.

---

[1]In fact, simulation of rotation by means of the Faraday effect, as mentioned in
Sec.2, requires already a description including birefringence.

[2]If we consider a ring laser instead of a passive ring, introduction of birefringence
leads to the so–called multifrequency laser gyro [2].

[3]A similar formalism describes wave propagation in a birefringent DFB structure,
a so–called Solc filter [9].

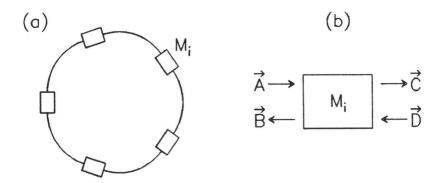

Fig.5   a) Generalized ring resonator. b) Optical element as transmission matrix.

Instead of giving a complete display of these 4–mode level crossings we will briefly dwell on their symmetry properties, restricting ourselves to rings with conservative transmission matrices M_i. These matrices M_i form the group U(2,2). The group U(2,2) has 16 generators, i.e. the birefringent optical ring resonator has at maximum 16 degrees of freedom, irrespective of the nature or the number of constituting ring elements M_i. (The value 16 corresponds to the number of independent elements of a 4×4 Hermitian matrix).

Level crossings in the birefringent optical ring should obey the general Dyson classification of eigenvalue crossings [24]. According to Dyson, realization of an eigenvalue crossing which corresponds to a doubling of the degeneracy requires control of N parameters, with $N = 2$, 3 or 5. For the birefringent optical ring, in special high–symmetry configurations $\{M_i\}$, the Hamiltonian is a 2×2 matrix with elements that are either real or complex or quaternion numbers. A simple count of independent matrix elements for these three cases then yields $N = 2, 3, 5$. Therefore, the optical ring allows explicit demonstration of the Dyson classification; the case $N = 5$ can only be realized in a birefringent ring[4].

For completeness we note that in the absence of birefringence the group of $\{M_i\}$ is U(1,1) in general and SU(1,1) if the Sagnac effect is absent. The latter group appears also elsewhere in optics, in descriptions of 2–photon squeezed states and of degenerate parametric amplifiers [25]. If backscattering is absent the group of $\{M_i\}$ is U(2) $\oplus$ U(2).

6. MODE–COUPLING IN A 2–MODE RING LASER

We consider now mode–coupling in an active ring, i.e. in a ring in which the cw

---

[4]The main application of the Dyson classification is in the statistical theory of spectra; in this sense there is, of course, no connection with the spectrum of a macroscopic (and therefore deterministic) optical ring.

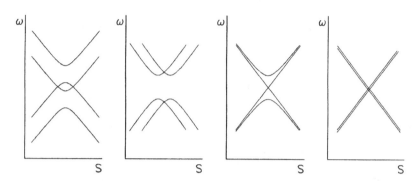

Fig. 6   Examples of level crossings of 4 optical modes coupled by backscattering and birefringence.

and *ccw* modes, coupled by backscattering, may oscillate due to insertion of a gain medium which overcomes the losses. Such a bidirectional ring laser with backscattering has been extensively investigated, due to its applicability as a Sagnac gyroscope [1–3] and also because of its interesting statistical properties [6–8]. The novel aspect as compared to the passive ring is that the *cw* and *ccw* modes have not only a linear coupling (conservative or dissipative), due to backscattering, but also a nonlinear one, due to gain saturation. We will restrict ourselves to cases where the linear coupling remains much stronger than the nonlinear one, $|R_{12}|_{\text{lin}} >> |R_{12}|_{\text{nl}}$ [5]. Using theoretical expressions [6] for $|R_{12}|_{\text{nl}}$ we estimate that for HeNe, dye and semiconductor ring lasers, under practical operating conditions $|R_{12}|_{\text{nl}} < 10^{-2} - 10^{-3}$. If linear coupling dominates, we may still use the normal modes of the passive ring as eigenmodes; the (weak) nonlinear coupling then introduces competition between these eigenmodes. The nature of the competition depends on the spectral broadening mechanism of the atomic transition which supplies the gain [3]. Numerical solution of the equations given in Ref. [3] leads to results as sketched in Fig.7. In the case of dissipative coupling (Fig.7a) both normal modes oscillate necessarily simultaneously, regardless of competition, since, as discussed in Sec.2, they are mutually nonorthogonal. As a consequence, the lockband (2*W*) of a ring laser gyro with purely dissipative linear coupling is independent of the nonlinear coupling; the lockband is completely determined by the characteristics of the passive ring (The nonlinear coupling *has* an effect in the sense that it leads to a nonsinusoidal character of the beat $\omega_+ - \omega_-$ if $|\Delta| > W$).

In the case of conservative linear coupling we must distinguish between homogeneous and Doppler broadening of the gain medium. If homogeneous broadening prevails, spatial hole burning occurs; therefore the valence and conduction band standing–wave normal modes (90^0–dephased) do not compete so

---

[5] In the opposite limit, $|R_{12}|_{\text{nl}} >> |R_{12}|_{\text{lin}}$, the problem belongs to the area of nonlinear wave dynamics [6].

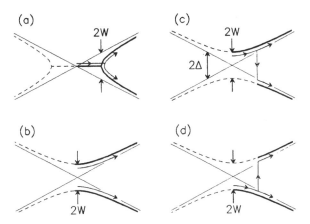

Fig.7   Effect of competition on normal–mode oscillation in a ring laser with backscattering. A heavy line indicates oscillation of the normal mode involved. a) Dissipative coupling. b) Conservative coupling, homogeneous gain medium. c) Conservative coupling, Doppler–broadened gain medium, tuning in red Doppler wing. d) Conservative coupling, Doppler–broadened gain medium, tuning in blue Doppler wing.

that they oscillate simultaneously (Fig.7b). Such a laser shows peculiar behavior, sometimes referred to as "oscillatory instability" [6,7]: the output changes from cw to ccw at a frequency $2W$, i.e. $I_{cw} = \cos^2 2Wt$ and $I_{ccw} = \sin^2 2Wt$ (see Fig.8).

If Doppler broadening dominates, spatial hole burning cannot occur, due to motional smearing, so that both standing waves compete for the same gain[6]. As a result, either the valence band mode or the conduction band mode oscillates, depending on whether the laser is tuned in the red or blue wing of the Doppler transition (Fig.7c,d). (Here we have assumed that the frequency–splitting is smaller than the residual homogeneous linewidth). For red wing tuning the conduction band mode is closer to gain maximum and therefore wins, whereas for blue wing tuning the valence band mode prevails. This single–frequency regime corresponds to "frequency–locked" operation; Sagnac beats are not present if the ring is set into rotation. If, however, the uncoupled–mode splitting Δ is increased, mode competition weakens and 2–mode oscillation, with a Sagnac related beat frequency $\omega_+ - \omega_-$, starts at some critical value of Δ (Fig.7c,d). Again, the beat is nonsinusoidal due to the nonlinear coupling. Thus "conservative" frequency locking of a ring laser gyro is clearly more complex than its "dissipative" counterpart (Fig.7a),

---

[6]Note that for counterpropagating *traveling* waves, again in a ring laser, the situation is reversed: in that case homogeneous broadening leads to competition whereas Doppler broadening does not.

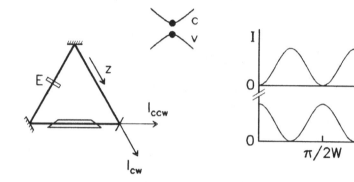

Fig.8   Time dependent output of homogeneously broadened ring laser with
            conservatively coupled *cw* and *ccw* modes (this case corresponds to
            Fig.7b). The laser oscillates simultaneously in valence and conduction
            band standing–wave normal modes.

since not only the linear but also the nonlinear coupling affects the lockband[7]. This
may lead, for instance, to hysteresis in the locking behavior.

In some treatments [1,7] of ring lasers with backscattering, *traveling* waves
(*cw,ccw*) are used as a basis, instead of standing waves (*val, cond*). The
normal–mode frequency splittings then appear in the time domain, e.g. in
time–correlation functions. Although formally equivalent, we feel that such an
approach is less conspicuous than the normal–mode approach presented here; the
contrast is like that between using momentum eigenstates and energy eigenstates
in quantum mechanics. For instance, π–jumps in the relative phase of *cw* and *ccw*
output beams [8] may now be interpreted as jumps from valence to conduction
band (or vice versa). Also, the oscillatory instability [6,7] mentioned above may
now be interpreted as a "quantum beat", due to simultaneous oscillation in valence
and conduction bands.

Finally, the reader should be aware that in several treatments of frequency
locking in ring laser gyros, in particular in the engineering literature, the
traveling–wave amplitudes a_1 and a_2 are assumed to be independent of time; also
nonlinear coupling is not always incorporated. Results of those treatments should
therefore be considered with care.

7. DYNAMIC EFFECTS IN OPTICAL AND ELECTRONIC BAND STRUCTURES

In this section we will explore analogies between optical and electronic band

---

[7]Note that frequency locking of electrical *LCR* circuits is also much simpler for
"resistive" than for "reactive" coupling [11].

structures, emphasizing the dynamic aspects. Let us first make the connection between the concepts of level crossing and band structure. Consider, as in Sec.2, the one–dimensional problem of two waves, with wave vectors k_1 and k_2, propagating in an (effectively) periodic medium with wave vector K. Coupling occurs not only for $k_1 = -k_2 = \frac{1}{2}K$, the case considered in Sec.2, but generally for $k_1 - k_2 = q\,K$. The manifold of level crossings associated with $q = \pm1, \pm2, \pm3...$ represents a band structure $\omega = \omega(S)$ or $E = E(S)$, where S is the dynamic variable. The generalized band structure shown in Fig.9 may thus refer to coupling of cw and ccw light waves, with different mode indices m, in an optical ring resonator. In this case, as discussed in Sec.2, the dynamic variable is Ω for a rotating ring or B for a ring with a Faraday element. Fig.9 may also refer to coupling of left– and right–traveling electronic waves in a periodic lattice of ions; the dynamic variable is then the electron wave vector k [26]. Another electronic implementation of Fig.9 deals with a small–capacitance tunnel junction (Josephson or normal) driven by a DC current source. The charge imbalance of the junction capacitance is quantized and the corresponding eigenstates are coupled by tunneling (of Cooper pairs and/or quasiparticles) [27–32]. In this case the dynamic variable is Q, the external charge which has flown from current source to tunnel junction.

Dynamic effects in band structures are of current interest, in particular in the electronic case [26–32]. Sufficiently fast variation of the dynamic variable S as a function of time may produce a nonadiabatic response in which the bandgap is crossed (Zener tunneling, see Fig.9). Sufficiently slow variation leads to adiabatic evolution along one of the bands; during this process the "group velocity" $(d\omega/dS)$ reverses its sign periodically so that a DC action leads to an AC response (Bloch oscillation, see Fig.9). The response is nonadiabatic or adiabatic, depending on whether $(2W)^2(d\Delta/dt)^{-1} << 1$ or $>> 1$, respectively [4,31,32]. Electronic Bloch oscillations are a hot topic in solid–state physics. They have never been directly observed so far; very recently indirect evidence has been reported [33]. This has motivated us to search for Bloch oscillations in the 1–dimensional photon band structure of an optical ring resonator [4]. Recently we have succeeded [22], thus demonstrating that Bloch oscillations can occur in classical–wave physics.

Naturally, this leads to the question: How useful is optical modeling of electronic phenomena? This question is also appropriate when dealing with optical modeling of 3–dimensional band structures [34] and of localization phenomena in random media [35].The justification of such optical modeling might be based on the fact that coupling of two modes can always be described by Eqs.(1,2). This is equivalent to the statement that the scalar wave equation of classical waves (e.g. light) reduces to the quantum mechanical Schrödinger equation if we impose the Slowly–Varying–Envelope–Approximation (SVEA). There are, nevertheless, important differences between photons and electrons, such as:

(i)   Photons are bosons and correspond to a classical field with a well–defined phase and amplitude. Electrons are fermions and cannot form such a coherent field.

(ii)   Photons carry no charge, contrary to electrons; as a consequence photons experience no s–wave scattering as electrons do. Another consequence is that photons do not interact with one another, whereas electrons do.

(iii)   Photons have a spin 1 whereas electrons have a spin $\frac{1}{2}$ (classically, the photon spin corresponds to the polarization vector).

Rigorously speaking, these differences make optical modeling of electronic

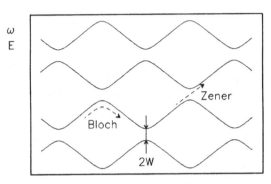

$$S= \Omega, \ B, \ k, \ Q, \ ...$$

Fig.9   Generalized band structure; dashed arrows indicate Bloch oscillation
         and Zener tunneling.

phenomena a futile excercise. In practice, however, the differences may be
mitigated due to a variety of reasons, such as:
(i)   The electronic wave retains some coherence if the number of electrons in
         the sample remains "small". The meaning of "small" depends on
         temperature and on whether we deal with a semiconductor or a metal
         [27]. (Of course, the electronic wave is fully coherent in the case of a
         superconductive system, where we deal with Cooper pairs, i.e. bosons).
(ii)   Electron–electron interaction may be largely screened in practical cases.
         Furthermore, it can possibly be simulated by the photon–photon
         interaction due to nonlinear–optical effects.
(iii)   In the paraxial–optical approximation the photon spin is effectively $\frac{1}{2}$
         since a Jones vector can be considered as a spinor.
    The overall conclusion is essentially the same as that which one may draw from
a discussion between Ehrenfest and Pauli [36,37], nearly 60 years ago, on the
analogy of electrons and photons: the analogy is purely heuristic but extremely
useful since it generates new ideas by cross fertilization. In this context it is
interesting to note that in present times the inspiration has flown more from
solid–state physics to optics than vice versa, thus mainly leading to new view
points and further development of already well–established optical subfields
[4,5,22,34,35]. However, we feel that in specific cases a well–designed optical
experiment may yield deeper insight into its solid–state physics counterpart. The
advantage of optical experiments, in particular of those on (macroscopic) ring
resonators, is that they allow precise control of all parameters, contrary to
experiments in solid–state physics. This applies in particular to the control of
dissipation, or inelastic processes, which tend to defeat coherent electronic effects
such as Bloch oscillations and localization [27–30,35].
    In the electronic case dissipation leads to conceptual problems, when it has to
be incorporated in a quantum mechanical theory [29,31,32], and also to
technological problems, when it has to be minimized in a specific solid–state
device. In optics, various kinds of dissipation are easily implemented. For example,
when modeling a (very small) capacitive Josephson junction, Ohmic losses and
losses associated with quasiparticle tunneling should be distinguished [29]. These

can be independently introduced in optics, using the birefringent optical ring discussed in Sec.4. In optics, dissipation may also be eliminated by (nearly) compensating unavoidable losses with an optical amplifier [22]. (This brings us very close to the ring laser of Sec.5). Presently we are preparing such experiments, in order to study the effect of dissipation on Bloch oscillations and Zener tunneling in the photon band structure. As explained above, we hope that these optical experiments will have a (heuristic) bearing on unresolved issues in the field of electronic Bloch oscillations and Zener tunneling.

ACKNOWLEDGMENT

Helpful discussions with R. Centeno Neelen, D. Lenstra, E.R. Eliel, H. de Lang, M. Kuś and F. Haake are gratefully acknowledged.

This work is part of the research program of the Foundation for Fundamental Research on Matter (FOM) and is supported financially by the Netherlands Organization for Scientific Research (NWO).

REFERENCES

[1] H.A. Haus, H. Statz and I.W. Smith, IEEE J. Quantum Electron. QE–21 (1985) 78.
[2] W.W. Chow, J. Gea–Banacloche, L.M. Pedrotti, V.E. Sanders, W. Schleich and M.O. Scully, Rev. Mod. Phys. 57 (1985) 61.
[3] L.N. Menegozzi and W.E. Lamb, Phys. Rev. A8 (1973) 2103.
[4] D. Lenstra, L.P.J. Kamp and W. van Haeringen, Opt. Commun. 60 (1986) 339.
[5] R.J.C. Spreeuw, J.P. Woerdman and D. Lenstra, Phys. Rev. Lett. 61 (1988) 318.
[6] D. Kühlke, Acta Phys. Pol. A61 (1982) 547 and references therein.
[7] W.R. Christian and L. Mandel, J. Opt. Soc. Am. B5 (1988) 1406 and references therein.
[8] W.R. Christian, T.H. Chyba, E.C. Gage and L. Mandel, Opt. Commun. 66 (1988) 238.
[9] P. Yeh, *Optical waves in layered media*, Wiley, New York (1988).
[10] H. Kogelnik and C.V. Shank, J. Appl. Phys. 43 (1972) 2327.
[11] A.B. Pippard, *The Physics of Vibration*, Vol. 1, Cambridge University Press, Cambridge (1978).
[12] G. Weinreich, J. Acoust. Soc. Am. 62 (1977); Scient. American (Jan. 1979) p. 94.
[13] D. Heitmann, N. Kroo, C. Schulz and Zs. Szentirmay, Phys. Rev. B35 (1987) 2660.
[14] H.A. Bethe, Phys. Rev. Lett. 56 (1986) 1305.
[15] P.W. Kabir, *The CP Puzzle*, Academic Press, New York (1968).
[16] C. Cohen–Tannoudji and S. Reynaud, J. Phys. B10 (1977) 365.
[17] M.L. Zimmerman, M.G. Littman, M.M. Kash and D. Kleppner, Phys. Rev. A20 (1979) 2251.
[18] H. Wieder and T.G. Eck, Phys. Rev. 153 (1967) 103.
[19] W. van de Water, D.R. Mariani and P.M. Koch, Phys. Rev. A30 (1984) 2399.
[20] P.G. Wolynes, J.Chem.Phys. 86 (1987) 1957.
[21] W.E. Lamb, Phys. Rev. 85 (1952) 259.

[22] R.J.C. Spreeuw, E.R. Eliel and J.P. Woerdman, Opt. Commun., in press.
[23] D. Lenstra and S.H.M. Geurten. Opt. Commun., in press.
[24] F.J. Dyson, J. Math. Phys. 3 (1982) 1199.
[25] C.C. Gerry, Phys. Rev. A39 (1989) 3204.
[26] L. Esaki and R. Tsu, IBM. J. Res. Develop. 14 (1970) 61.
[27] M. Büttiker, Y. Imry and R. Landauer, Phys. Lett. 96A (1983) 365.
[28] D.V. Averin and K.K. Likharev, J. Low Temp. Phys. 62 (1986) 345.
[29] F. Guinea and G. Schön, J. Low Temp. Phys. 69 (1987) 219.
[30] E. Ben–Jacob, Y. Gefen, K. Mullen and Z. Schuss, Phys. Rev. B37 (1988) 7401.
[31] K. Mullen, E. Ben–Jacob, Y. Gefen and Z. Schuss, Phys. Rev. Lett. 62 (1989) 2543.
[32] P. Ao and J. Rammer, Phys. Rev. Lett. 62 (1989) 3004.
[33] L.J. Geerligs, M. Peters, L.E.M. de Groot, A. Verbruggen and J.E. Mooij, Phys. Rev. Lett. 63 (1989) 326.
[34] E. Yablonovitch, Phys. Rev. Lett. 58 (1987) 2059.
[35] M.P. van Albada and A. Lagendijk, Phys. Rev. Lett. 55 (1985) 2692.
[36] P. Ehrenfest, Z. Physik 78 (1932) 555.
[37] W. Pauli, Z. Physik 80 (1933) 573.

J.P. Woerdman and *R.J.C. Spreeuw* are with the University of Leiden, Huygens Laboratory, P.O. Box 9504, 2300 RA Leiden, The Netherlands

BERRY'S PHASES IN OPTICS

Raymond Y. Chiao

Berry's discovery of a topological phase in quantum mechanics has led to a unified view of many seemingly disconnected topological phenomena in physics, both at the quantum and classical levels. In analogy with the wavefunction of an electron, the wavefunction of a photon can also acquire this topological phase. Here we review four recent optical manifestations of this Aharonov–Bohm–like phase. A distinction is made between classical and nonclassical Berry's phases in optics. One example of a nonclassical phase is the one associated with a cycle of Von Neumann projections. Other nonclassical phases will be discussed. A group–theoretical analysis of classical Berry's phases for Maxwell's equations expressed as a Schrödinger–like equation for a six–component spinor is introduced.

1. INTRODUCTION

Analogies between wave propagation in optics and in electronics have been fruitful ones. In fact, it was this analogy which first led Schrödinger to his equation. As with Newton's equation of motion, and with Maxwell's equations, Schrödinger's equation is a *differential* equation. Therefore it expresses *local* relationships between physical quantities. However, there exist important *global* and *topological* properties of the solutions of Schrödinger's equation, and of Maxwell's equations, which are difficult to recognize. These have recently been brought to light, especially following Berry's discovery of a geometrical, or, as we shall call it, *topological* phase in the quantum adiabatic theorem. This phase is very similar to the Aharonov–Bohm phase which an electron acquires after encircling a solenoid. In fact, one can view the Aharonov–Bohm phase as a special case of Berry's phase. The question naturally arises: Do such phases also exist for the photon? Here we review progress towards answering this question.

There have been four recent manifestations of Berry's phases in optics. All these phases can be thought of as anholonomies. An *anholonomy* can arise whenever a system is processed through a sequence of changes such that it is returned to its original state; most of the physical variables characterizing the system obviously return to their original values, but surprisingly certain ones may *not*. An important example mentioned above is the *phase* anholonomy which an electron exhibits when it returns to its original state after encircling a solenoid in the Aharonov–Bohm effect. Another important example is the *age* anholonomy which a travelling twin experiences relative to the nontravelling twin when he returns to the Earth after a rocket trip in the twin paradox. When light is cycled through a sequence of states such that it returns to its original state, it can also acquire physical anholonomies at both the classical and quantum levels.

The phase anholonomy which Berry [1] discovered in the quantum adiabatic theorem can be expressed as an extra phase factor $\exp(i\gamma_n(C))$ which the

151

W. van Haeringen and D. Lenstra (eds.), Analogies in Optics and Micro Electronics, 151–161.
© 1990 *Kluwer Academic Publishers. Printed in the Netherlands.*

wavefunction can acquire after a cycle C in the parameter space of the Hamiltonian $H(R)$, where R denotes some slowly varying parameters which return to their starting values. This *topological* factor is accumulated by the wavefunction in addition to the usual *dynamical* phase factor, $\exp(-i\int E_n dt/\hbar)$. Explicitly, if $H(R)|n;R> = E_n(R)|n;R>$, then Berry's phase is

$$\gamma_n(C) = \oint_C A_{\text{eff}} \cdot dR, \tag{1}$$

where $A_{\text{eff}} = i<n;R|\nabla_R|n;R>$. This phase possesses the important property of *local gauge invariance*, viz., under the transformation

$$A_{\text{eff}} \to A_{\text{eff}} + \nabla_R \chi(R), \tag{2}$$

where $\chi(R)$ is any single–valued function of R, the phase $\gamma_n(C)$, and hence the phase factor $\exp(i\gamma_n(C))$, is invariant. This phase shift is physical, since it can be observed as a fringe shift in interference experiments. However, this phase is identically zero except in topologically nontrivial circumstances, like in the Aharonov–Bohm effect.

In an important generalization, Aharonov and Anandan [1] removed the adiabatic restriction. In so doing, they replaced the notion of *parameter space* by the notion of *state space*, which is the projective space of rays in Hilbert space. In (1), one now reinterprets R as coordinates on the state space manifold, rather than as coordinates on the parameter space manifold.

The phase factor associated with (1) is dependent on the path or past history of the system. Therefore, it is another example of Yang's "nonintegrable phase factor" in physics, and hence leads to gauge phenomena. Here, however, these gauge phenomena do not occur in spacetime, as do the ones associated with the electromagnetic, weak, and strong interactions, but rather occur as *effective* gauge fields in state space. Although the latter seem nowhere as important as the former, these effective gauge phenomena serve as convenient analogs for studying the spacetime ones, and may shed some light on their origin.

2. SPIN REDIRECTION BERRY'S PHASE

The first recent optical manifestation of Berry's phase is the *angle* anholonomy which occurs when linearly polarized light enters a single–mode isotropic fiber which is wound into the shape of a helix [2]. One can arrange it so that the light returns to its original state in most of its variables, e.g., its frequency, its direction of propagation, and its state of polarization. However, the *axis* of linear polarization of this light does *not* return to its original direction, when referred to the axis of polarization at the entrance to the fiber. This can be understood at a classical level as arising from the *parallel transport* of the electric field vector of the light along the fiber [3].

A simple mechanical analogy helps us to understand this phenomenon [4]: Let one's extended right arm represent the instantaneous direction of propagation k of the light inside the fiber. Now extend the thumb such that it is always perpendicular to one's arm. The direction of the thumb represents the electric field vector E of the light, which is always perpendicular to k. Let us start with our extended right arm pointing straight in front of us, and our thumb pointing upwards. Let this represent the original state of the light entering the fiber.

Now move one's arm upward toward the vertical, *keeping the thumb parallel to itself at all times*. This represents the changing direction of light propagation inside the fiber, with the electric field vector undergoing parallel transport during this process. The arm thus sweeps out an arc in a vertical plane. After one's arm reaches the vertical position, move it downward along a perpendicular vertical arc, so that the arm extends to one's right. The thumb now points backward. Complete the cycle by moving one's arm in a horizontal arc, so that the *arm* is brought back to its starting position pointing forward. Surprisingly the *thumb* does *not* end up pointing in its starting direction. It now points towards the right, not upward, as if it had been rotated by 90° around the axis of the arm. *Locally*, at each point in the motion, there has been no rotation of the thumb around the axis of the arm, since it has been transported parallel to itself; *globally*, however, there exists such an effective rotation.

The above parallel transport process, apart from a difference of spaces, is identical to the one in general relativity, where it embodies *the principle of equivalence*. Here, however, we do not invoke parallel transport as a *principle*, but rather it should be explicitly derivable from the underlying dynamical equations (i.e. Maxwell's equations) along with the symmetry (i.e. local isotropy) of the medium [2]. In fact, Berry has done this recently [25]. Here dynamics precedes geometry, rather than geometry preceding dynamics.

The net result of this cycle of changes in direction of light is an *angle anholonomy*: There is a net *rotation* of the plane of polarization of the light by 90° in the above example. In general, the angle of rotation of the E vector (the thumb) is equal to the solid angle subtended by the closed curve swept out by the k vector (the arm). This is a consequence of the Gauss–Bennet theorem, which states that the sum of the three interior angles of a spherical triangle is 180° plus the solid angle enclosed by that triangle.

This angle anholonomy can also be thought of as an *effective* optical activity induced by the helical path of the light. For instance, one can think of the helically wound single–mode optical fiber, composed of an isotropic glass medium without any intrinsic optical activity, as *if* it were an optically active medium. However, in contrast to ordinary optical activity, the magnitude of the geometrically induced optical rotation inside the helically wound fiber does not depend on the wavelength of the light, a point we have verified experimentally [5].

Associated with this *angle* anholonomy is a *phase* anholonomy for circularly polarized light. The spin vector of a photon in a beam of circularly polarized light, after a cycle of changes in its direction, is rotated around its axis by this angle, thus resulting in a phase shift between opposite senses of circularly polarized light [2,6], which has been observed in a nonplanar Mach–Zehnder interferometer [7]. This phase shift does not originate in the optical path length difference in the two arms of the interferometer, but rather in the difference in their handedness, a topological effect. The dynamical phase, which is proportional to the optical path length, cancels out in both the optical fiber experiment [2] and the Mach–Zehnder experiment [7]. The former experiment was *adiabatic*; the latter *nonadiabatic*.

State spaces are the arenas in which these topological and geometrical phenomena associated with Berry's phases occur. A cycle of changes in the system corresponds to a closed curve in its state space, whose geometry and topology determine Berry's phases. Associated with the above example is a state space consisting of the surface of a sphere whose points correspond to all possible directions of the photon spin. This surface constitutes the state space manifold.

Note that the *curvature* of this manifold is not zero. We shall call the associated Berry's phase *the spin redirection phase*. This phase is given by [2]

$$\gamma(C) = -\sigma\Omega(C), \tag{3}$$

where C is a circuit on the sphere, $\sigma = \pm 1$ is the helicity of the photon, and $\Omega(C)$ is the solid angle subtended by C with respect to the center of the sphere. This phase is formally identical to the Aharonov–Bohm phase which an electron picks up after an analogous circuit on a sphere centered on a Dirac monopole.

3. PANCHARATNAM'S PHASE

In a second example of Berry's phase, the state space is the Poincaré sphere, which describes all possible polarization states of light. A cycle of changes in polarization states, with the direction of the light kept fixed, corresponds to a closed curve C on the Poincaré sphere. The associated Berry's phase is "Pancharatnam's phase" [8]. This phase $\gamma(C)$ is given by

$$\gamma(C) = -\Omega(C)/2, \tag{4}$$

where C is the circuit on the Poincaré sphere. Note that the fraction of one–half distinguishes this phase from that in (3).

We have studied the mutual influence of these two Berry's phases by means of the nonplanar Mach–Zehnder interferometer which we constructed in our earlier work [7,9]. If we introduce a modified state space, namely, a generalized Poincaré sphere [10], then the above two phases are additive. The generalized Poincaré sphere defines the polarization state of the light with respect to space–fixed axes, not the propagation direction of the light, as in the usual Poincaré sphere. Hence the generalized sphere refer to the *angular momentum* of the light, whereas the usual sphere refers to the *helicity* of the light. Since it is angular momentum, not helicity, which is exchanged between the light and the optical components (this is the source of these anholonomies), the generalized Poincaré sphere plays the role of the state space here.

4. SQUEEZED–STATES BERRY'S PHASE

The generators of squeezed states of light are $K_1 = -i(aa - a\dagger a\dagger)/4$ and $K_2 = (aa + a\dagger a\dagger)/4$, where $[a, a\dagger] = 1$ [11]. These along with $J_3 = (aa\dagger + a\dagger a)/4$ satisfy the commutation relations

$$[K_1, K_2] = -iJ_3, \qquad [K_2, J_3] = iK_1, \qquad [J_3, K_1] = iK_2, \tag{5}$$

which generate the group SU(1,1) [12]. This is isomorphic to the Lorentz group in two spatial dimensions [13]. Apart from a sign, these commutation relations are identical to those of the three angular momentum components which generate the group SU(2). The group manifold for SU(2) is a sphere. Starting with an arbitrary state vector, the generators can transport this state to all other possible state vectors of the system. Hence the group manifold is identical to the state space manifold; the state space for SU(2) is therefore a sphere. For example, in the case of Pancharatnam's phase, it is the Poincaré sphere. Likewise, one can construct the state space for squeezed states by letting the elements of SU(1,1) act on any given ray in state space for squeezed states, e.g., the vacuum state. The group manifold for SU(1,1) can be viewed as a *pseudosphere* in a Minkowski space–time with two space–like dimensions and one time–like dimension. This group manifold is identical to the state space manifold for squeezed states. The resulting Berry's phase is therefore formally identical to (4), when viewed in Minkowski space.

Alternatively, one can characterize the state space of squeezed states of light by three *real* numbers k_1, k_2, k_3 satisfying the relationships [13]

$$S_n...S_2 S_1 J_3 S_1\dagger\ S_2\dagger\ ...S_n\dagger = k_1 K_1 + k_2 K_2 + k_3 J_3, \tag{6}$$

$$-k_1{}^2 - k_2{}^2 + k_3{}^2 = 1, \tag{7}$$

where a sequence of squeezings associated with unitary operators $S_1, ... S_n$ takes J_3 to $k_1 K_1 + k_2 K_2 + k_3 J_3$. Here each $S_i = \exp\{ir_i(K_1\cos\theta_i + K_2\sin\theta_i)\}$, where r_i is the ith squeezing parameter and θ_i is the ith phase of squeezing. The state space for squeezed states in this view is a *hyperboloid* of revolution described by (7). A specific example for Berry's phase is that for a "square circuit", in which light is squeezed in one quadrature and then in an orthogonal quadrature, and then compared with the same light squeezed in the opposite order. Experimentally, this could be implemented by means of degenerate parametric amplifiers in an interferometer circuit, in which the phase and amplitude of the pump is appropriately adjusted from amplifier to amplifier. The standard Pauli matrix representations of the Lorentz–group generators $K_1 = i\sigma_z/2$, $K_2 = i\sigma_x/2$ and $J_3 = \sigma_y/2$ were used in calculating this phase [13].

An electronic analogy for the above cycle of squeezings of light is the following: Place an electron on a frictionless table. Initially, the electron is at rest at the origin of a Cartesian coordinate system, with x and y axes aligned with respect to the edges of the table, and the z axis perpendicular to the tabletop. Let its spin point in z direction. Now Lorentz boost the electron in the x direction, and follow this by a boost in the y direction. Then boost it in the $-x$ direction, and follow this by a boost in the $-y$ direction, such that it comes back to rest. After this cycle of Lorentz boosts, the electron experiences an angle anholonomy: it comes back to its orginal state of rest rotated by the Wigner rotation angle ϕ around the z axis.

The associated phase factor arises from the net rotation operator $\exp(i\phi J_3)$. Thus the resulting phase factor for a Fock state $|n>$ so cycled is $\exp\{i\phi(n+\tfrac{1}{2})/2\}$. For a coherent state $|\alpha>$ so cycled, the result is that $|\alpha> \rightarrow \exp(i\phi/4) \times |\alpha \exp(i\phi/2)>$. One interesting feature of these results is that when a Fock state $|n>$ is processed through such a cycle of squeezings, the resulting Berry's phase is proportional to $n + 1/2$, implying that it is proportional to the intensity of the light. However, when a coherent state is taken through the same cycle, there is no such proportionality. The Fock state Berry's phase is therefore nonclassical (see below).

5. THE KITANO–YABUZAKI PHASE

Kitano and Yabuzaki [14] noticed that there is another optical situation where a Lorentz–group Berry's phase appears. They found that if one cycles light through a sequence of *partial* linearly polarized states, the light acquires a phase which is formally identical to the above squeezed–states Berry's phase.

A partial linear polarizer can be represented by the 2×2 matrix

$$S = \mathrm{diag}(t_x,t_y) = t_a \,\mathrm{diag}(\kappa,\kappa^{-1}), \tag{8}$$

where $t_x(t_y)$ is the transmission coefficient for the $x(y)$ component of the field, $t_a = (t_x t_y)^{1/2}$ and $\kappa = (t_x/t_y)^{1/2}$. The role of the squeezing parameter r is played here by $-2ln\kappa$, which is a partial linear polarization parameter. The standard Pauli matrix representations of the Lorentz–group generators used above work here as well. Here the physical meaning of these generators is that they generate partial linear polarization along the x axis, partial linear polarization along an axis $45°$ to the x axis, and a rotation around the beam axis, respectively.

Associated with this phase anholonomy is an angle anholonomy, in which there is an extra rotation of the axis of partial linear polarization after it has passed through a sequence of four partial linear polarizers. This extra rotation angle is analogous to the Wigner rotation angle of the Lorentz group, and to the Thomas precession angle of electron spin. They chose the axes of these polarizers to be at $0°$, $45°$, $90°$ and $135°$, and observed this angle anholonomy by using tilted glass plates as partial linear polarizers.

This phase is an example of the extension of Berry's phase to statistical mechanics, since partially polarized light is a prototype of *mixed* states. States are then described by the density matrix instead of the wavefunction [15]. It would be interesting to generalize Kitano and Yabuzaki's result to arbitary states of partial polarization of light. Another interesting avenue of exploration would be to look at anholonomies resulting from a cycle in states of partially coherent light.

6. NONCLASSICAL BERRY'S PHASES IN OPTICS

Are these phases classical or quantal? In all of the cases mentioned above, except for the Fock state Berry's phase resulting from squeezings, we can view them either as phase shifts of a classical electromagnetic *wave* or as phase shifts of the *wavefunction* for the photon. Even the coherent state Berry's phase which results from a cycle of sqeezings of a coherent state can be viewed classically, e.g., even a *microwave* degenerate paramp interferometer circuit will exhibit this phase shift [13]. However, there are two situations in optics where purely quantal, i.e., nonclassical, Berry's phases appear. The first arises when the light is in a *nonclassical state* that does not possess any correspondence principle limit in which it turns into a classical wave, unlike a coherent state. Typical nonclassical states are Fock states and squeezed states. A good example is the Fock state Berry's phase resulting from a cycle of squeezings. The second purely quantal Berry's phase arises when the *configuration space* of two or more photons makes its appearance as the state space of the system.

As another example of the first nonclassical situation, consider the Berry's phase arising from quantum measurements [16]. Consider a Stern–Gerlach experiment in which a sequence of three filterings is performed in the x, y, and z directions on a beam of spins, followed by a fourth filtering in the x direction. At each step, let us filter out the $m_s = -1/2$ component in the beam by absorbing these particles, and pass the $m_s = +1/2$ component onto the next filter. Each filtering prepares the system in the quantum state $|n; m_s=+1/2>$, where n denotes the spatial quantization axis of the apparatus. Accompanying each filtering process is a discontinuous "collapse of the wavefunction", i.e., a discontinuous projection of the incoming ket onto the eigenkets of the measuring apparatus. This process can be described by the Von Neumann projection postulate

$$|\psi> \ \rightarrow \ |\,n;\ m_s \,=\, +1/2><n;m_s=+1/2\,|\,\psi>/\|<n;\ m_s=+1/2\,|\,\psi>\|. \qquad (9)$$

After the first filter, the system is in a state $|\,x\,;\,+1/2>$, and after the fourth filter, $\exp(i\gamma(C))|\,x\,;\,+1/2>$. Here C is a triangular circuit on the sphere of spin directions consisting of geodesics joining the three points corresponding to the x, y and z spin directions. The solid angle is $\pi/2$, and hence $\gamma(C)=-\pi/4$. This phase can be observed in interference.

We can replace spin 1/2 particles by any two–state systems. Photons, which possess two polarization states, and which exhibit interference readily, are a good choice. We can then replace the Stern–Gerlach apparati by polarizers. Corresponding to the above example, we can use two linear polarizers with 45° between their axes, followed by a circular polarizer and another linear polarizer oriented parallel to the first polarizer. Let these filters be dichroic. The state space is again the Poincaré sphere. The resulting Berry's phase is *nonclassical* if we prepare the photons in $|\,n{=}1>$ Fock states [17]. This experiment would answer the following questions: Does the phase of the wavefunction survive the "collapse of the wavefunction"? Is this surviving phase geometrical or topological? Are the geodesic constructions used in calculating this phase valid? Is the Von Neumann projection postulate correct?

As an example of the second nonclassical situation, consider the two–particle wavefunction $\psi(r_1,r_2;t)$ defined on the configuration space $(r_1,r_2;t)$ of the particles, which one can think of as a state space of the system. If these particles undergo a sequence of changes such that they return to their original state, then in general

$$\psi(r_1,r_2;t) \ \rightarrow \ \exp(i\gamma(C))\ \psi(r_1,r_2;t), \qquad (10)$$

where $\gamma(C)$ is a phase which in general depends on their history C, represented by a circuit in configuration space. An important special case arises when these two particles are identical, and their positions are interchanged by a transposition. The resulting phase in three dimensions gives rise to Bose or Fermi statistics, a nonclassical result. Now restrict the configuration space coordinates $(r_1,r_2;t)$ to angular variables $(\phi_1,\phi_2;t)$. Since the configuration space $(\phi_1,\phi_2;t)$ is doubly periodic with periodicity 2π in both of the two variables ϕ_1 and ϕ_2, the state space in this case is a torus, whose geometry and topology is nontrivial. For example, consider a two–photon state prepared in an Einstein–Poldolsky–Rosen "singlet" combination of linear polarization states [18]

$$|\psi> \ = \ 2^{-1/2}\{\,|\,x>_1|\,y>_2 - |\,y>_1|\,x>_2\}. \qquad (11)$$

The wavefunction $\psi(\phi_1,\phi_2;t) \ = \ <\phi_1,\phi_2;t|\,\psi>$ is the probability amplitude of finding the two photons linearly polarized along the ϕ_1 and ϕ_2 directions simultaneously, i.e., in coincidence at time t. Now place in the path of photon 1 an optically active medium which rotates its plane of polarization through 180°. The polarization of photon 1 is thus turned through space by 180°, which results in the sign change

$$\psi(\phi_1,\phi_2;t) \ \rightarrow \ \exp(i\pi)\psi(\phi_1,\phi_2;t) = -\psi(\phi_1,\phi_2;t). \qquad (12)$$

This phase is observable in two–particle interferometry [19].

7. BERRY'S PHASES FOR MAXWELL'S EQUATIONS

As we have seen above, many Berry's phases have been uncovered in optics. Some admit classical explanations, while others do not. Berry's phases also occur in nonlinear optics [20]. Questions which naturally arise include: How many Berry's phases are there in optics? How are they related to each other? How can we unify them? Instead of answering these questions in their full generality, let us restrict ourselves to the following question: How many Berry's phases are there for Maxwell's equations?

In order to answer this question, let us recast Maxwell's equations into a Schrödinger–like form, so that we can apply the existing analyses of Berry's phase to them. Note that $\hbar$ does not appear in these equations, so that the following analysis is entirely on the classical level. However, the existing analyses of Berry's phase are still applicable, since they apply to all Schrödinger–like equations, i.e., partial differential equations which involve i times the first time derivative.

Let us begin with the vacuum Maxwell's equations. *In vacuo*, the time–dependent Maxwell's equations are (with $c=1$):

$$\text{curl } E = -\partial B/\partial t, \tag{13}$$

$$\text{curl } B = +\partial E/\partial t. \tag{14}$$

Multiplying the latter equation by $\pm i$ and adding it to the former equation, we obtain

$$\text{curl } (E \pm iB) = \pm i\,\partial/\partial t\,(E \pm iB)\,. \tag{15}$$

Let ψ be the six–component spinor,

$$\psi = \text{col}(E{+}iB, E{-}iB) = \text{col}(\psi^{(+)}, \psi^{(-)}), \tag{16}$$

where "col" is short for "column vector". We obtain the Schrödinger–like equation,

$$i\,\partial\psi/\partial t = H\psi, \tag{17}$$

where the Hamiltonian H is given by the 6×6 Hermitian matrix operator

$$\begin{bmatrix} (\text{curl}) & (0) \\ (0) & -(\text{curl}) \end{bmatrix}$$

where the round parentheses denote 3×3 submatrices. The resulting Schrödinger–like evolution of ψ is unitary.

Further conditions on the solutions are placed by the other two time–independent Maxwell's equations

$$\text{div } E = 0 \tag{18}$$

$$\text{div } B = 0, \tag{19}$$

which can be combined to give the transversality constraint,

$$\text{div } \psi = 0. \tag{20}$$

In order to include the effect of optical components like mirrors, quarter–wave plates, etc., we note that these devices have only two ports, i.e., an input and an output port. Let us assume for the moment that these elements are lossless, so that input and output vacuum solutions are connected by unitary matrices. Since the propagation in the vacuum between optical elements is also unitary, we can completely characterize the present problem by extracting the Berry's phases which result from a product of unitary matrices that return the state of the light beam back to its starting point, so that interference between the initial and final states becomes possible. Therefore, Jordan's analysis of Berry's phases based on closed loop sequences of unitary transformations, applies [21]. In particular, for Maxwell's equations in spinor form, (16) and (17), the most general possible unitary matrices which transform transverse plane waves into transverse plane waves are the elements of the subgroup of the group U(6) which satisfy the transversality constraint (20). The spin redirection phase and Pancharatnam's phase should emerge from this general analysis, as well as the Sagnac phase, if one includes moving mirrors.

Once Maxwell's equations are cast into a Schrödinger–like form, the analysis of Aharonov and Anandan [1], when generalized to six–component spinors, also applies. Although it is at present unclear how all of the smaller state spaces emerge in this general picture, it is easy to see how the Poincaré sphere emerges [9]. Since the light beams used in our experiments are essentially plane wave states, or, more precisely, Gaussian beams, the infinite dimensional Hilbert space is reduced to six complex dimensions. Note that the optical components in a typical interferometer, such as mirrors, quarter–wave plates, etc., do not alter the Gaussian functional form of these beams. In other words, Gaussian beams are always transformed into Gaussian beams. Let us further restrict our attention to the case in which the direction of the light beam is not changed by the insertion of optical components, such as quarter–wave plates. If we yet further restrict our attention to the case in which the two halves of the six–component spinor $\psi^{(+)}$ and $\psi^{(-)}$ are not mixed together by the optical components (the significance of this will be discussed below), the dimensionality of Hilbert space is reduced to three complex dimensions. Then the transversality condition, (20), reduces the spinors to two complex dimensions, whose state space Berry has shown is equivalent to the Poincaré sphere [22]. This is not surprising, since by the Feynman–Vernon–Hellwarth theorem any two–state problem can be reduced to an analogous spin 1/2 one, for which the state space is always a sphere.

Transformations between $\psi^{(+)}$ and $\psi^{(-)}$ correspond physically to optical components which produce optical phase conjugation, as can be seen by inspection of (16). Cycles of such transformations which can generate anholonomies have not yet been studied. Hence the restriction mentioned above to cases in which the two halves of the six–component spinor $\psi^{(+)}$ and $\psi^{(-)}$ are not mixed together by optical components, means that one must exclude optical phase conjugators from the system.

In order to calculate the phase anholonomy in the case of discrete optical components, one should use geodesics to connect points in state space. Samuel and Bhandari [23] have generalized Pancharatnam's geodesic constructions on the Poincaré sphere in optics [8], to all Schrödinger evolutions in physics. Since Maxwell's equations have been cast into a Schrödinger–like form, these geodesic constructions also apply to all of the other optical Berry's phases as well, including those in Refs. 7 and 9.

The above discussion suggests a general group–theoretical approach to the problem. Consider situations in which each optical component of an optical system transforms one solution of the vacuum Maxwell's equations into another. Each optical component produces a particular transformation within state space, e.g., a mirror produces a rotation on the sphere of all possible spin directions. Each transformation thus constitutes an element of a transitive group, whose elements transform a ray in state space into all other possible rays. One can therefore construct the state space by letting the transitive group act on any given ray. Hence the group manifold is identical to the state space manifold. The geometry and the topology of this manifold determine all of the physical anholonomies. This manifold is a base space above which one can erect various kinds of fiber bundles; a cycle of transformations then generates what is called in mathematics *the holonomy group*. For Maxwell's equations, the holonomy group is nonabelian, and the resulting effective gauge field is also nonabelian in general. The associated physical anholonomies need not be merely *phase* anholonomies, but will in general be *internal rotation* anholonomies, analogous to the ones discussed by Wilczek and Zee [24]. Further work is being pursued along these lines.

8. CONCLUSION

Analogies between wave propagation in optics and microelectronics have been fruitful. In particular, through the analogy between Schrödinger and Maxwell wave propagations, combined with Berry's prescription for finding topological phases, various topological phenomena in optics have been found. As a result, a rich variety of Aharonov–Bohm–like phenomena in optics have been uncovered. In addition to classical phenomena, there are also nonclassical ones. A unified description of these phenomena in group–theoretical terms seems to be emerging.

ACKNOWLEDGMENTS

I would like to thank all my collaborators for their contributions to this work, and J.C. Garrison, H.L. Morrison, and E.H. Wichman, for helpful discussions on group theory. This work was supported by the National Science Foundation under grant ECS 86–13773, and by the Office of Naval Research under contract N00014–88–K–0126.

REFERENCES

[1]   M.V. Berry, Proc. Roy. Soc. (London) A392 (1984) 45; J.H. Hannay, J. Phys. A18 (1985) 221; Y. Aharonov and J. Anandan, Phys. Rev. Lett. 58 (1987) 1593.
[2]   R.Y. Chiao and Y.S. Su, Phys. Rev. Lett. 57 (1986) 933; A. Tomita and R. Y. Chiao, *ibid.* 57 (1986) 937; R.Y. Chiao, Nucl. Phys. B (Proc. Suppl.) 6 (1989) 298.
[3]   J.N. Ross, Opt. Quantum Electron. 16 (1984) 445; F.D.M. Haldane, Opt. Lett. 11 (1986) 730.
[4]   R. Cherry, Sci. Am. 260(3) (1989) 9; I thank C. Wohl for pointing this analogy out to me.
[5]   H. Jiao, A. Tomita and R.Y. Chiao (unpublished).

[6]   J. Anandan and L. Stodolsky, Phys. Rev. D35 (1987) 2597; T.F. Jordan, J. Math. Phys. 28 (1987) 1759; *ibid.* 29 (1988) 2042.

[7]   R.Y. Chiao, A. Antaramian, K.M. Ganga, H.Jiao, S.R. Wilkinson and H. Nathel, Phys. Rev. Lett. 60 (1988) 1214.

[8]   S. Pancharatnam, Proc. Ind. Acad. Sci. A44 (1956) 247; R. Bhandari and J. Samuel, Phys. Rev. Lett. 60 (1989) 1210; T.H. Chyba, L.J. Wang, L. Mandel, and R. Simon, Opt.Lett. 13 (1988) 562; R .D. Simon, H.J. Kimble and E.C.G. Sudarshan, Phys. Rev. Lett. 61 (1988) 19; R. Bhandari, Phys. Lett. A133 (1988) 1; R. Bandhari discusses Pancharatnam's phase in detail in Chap.5 of this book.

[9]   H. Jiao, S.R. Wilkonson, R.Y. Chiao, and H. Nathel, Phys. Rev. A39 (1989) 3475.

[10]   W.R. Tompkin, M.S. Malcuit, R.W. Boyd, and R.Y. Chiao (to be published).

[11]   See the special issue on squeezed states in J. Opt. Soc. Am. B4 (1987) 1450.

[12]   R. Gilmore, *Lie Groups, Lie Algebras, and Some of their Applications* (John Wiley & Sons, New York, 1974).

[13]   R.Y. Chiao and T.F. Jordan, Phys. Lett. A132 (1988) 77; R.Y. Chiao, Nucl. Phys. B (Proc. Suppl.) 6 (1989) 327; C.C. Gerry, Phys. Rev. A39 (1989) 3204.

[14]   M.K. Kitano and T. Yabuzaki (preprint).

[15]   D. Suter, K.T. Mueller, and A. Pines, Phys. Rev. Lett. 60 (1988) 1218; A. Uhlmann, Rep. Math. Phys. 24 (1986) 229; J. Anandan, Phys. Lett. A133 (1988) 171; L. Dabrowski and H. Grosse (preprint).

[16]   J. Samuel and R. Bhandari, Phys. Rev. Lett. 60 (1988) 2339.

[17]   D.C. Burnham and D.L. Weinberg, Phys. Rev. Lett. 25 (1970) 84; R. Ghosh and L. Mandel, *ibid.* 59 (1987) 1903; C.K. Hong, Z.Y. Ou, and L. Mandel, *ibid.* 59 (1987) 2044; Z.Y. Ou and L. Mandel, *ibid.* 61 (1988) 50; Z.Y. Ou and L. Mandel, *ibid.* 61 (1988) 54; P. Grangier, A. Aspect, and G. Roger, Europhys. Lett. 1 (1986) 173.

[18]   C.O. Alley and Y.H. Shih, in *Proceedings of the Second International Symposium on Foundations of Quantum Mechanics in the Light of New Technology*, eds. M. Namiki *et al.* (Physical Society of Japan, Tokyo, 1986), p. 47; Y.H. Shih and C.O. Alley, Phys. Rev. Lett. 61 (1988) 2921.

[19]   J.D. Franson, Phys. Rev. Lett. 62 (1989) 2205; M.A. Horne, A. Shimony, and A. Zeilinger, *ibid.* 62 (1989) 2209.

[20]   J.C. Garrison and R.Y. Chiao, Phys. Rev. Lett. 60 (1988) 165.

[21]   T.F. Jordan, J. Math. Phys. 29 (1988) 2042.

[22]   M.V. Berry, J. Mod. Opt. 34 (1987) 1401.

[23]   J. Samuel and R. Bhandari, Phys. Rev. Lett. 60 (1988) 2339.

[24]   F. Wilczek and A. Zee, Phys. Rev. Lett. 52 (1984) 2111.

[25]   M.V. Berry, *Quantum Adiabatic Anholonomy*, Lectures at the Ferrara School of Theoretical Physics (June 1989), to be published.

R.Y. Chiao is with the University of California, Department of Physics, Berkeley, CA 94720, USA.

PART III

COHERENT ELECTRONICS

AHARONOV–BOHM OSCILLATIONS AND NON–LOCAL ELECTRONIC CONDUCTION

C. Van Haesendonck

In a mesoscopic (size smaller than the electron phase coherence length $L_\phi \sim 1\mu m$) disordered metal film or Si–MOSFET inversion layer, the ensemble averaging of the interference between scattered electron waves is incomplete. The tuning of the electron phase by the Aharonov–Bohm effect causes sample–specific magnetoconductance fluctuations with rms amplitude $G_{un} = e^2/h$. For a mesoscopic loop, the fluctuations are replaced by oscillations with flux period h/e. In the macroscopic regime, only the interference in the backscattering direction survives and causes the weak electron localization. In a long cylinder (diameter $\sim L_\phi$), the coherent backscattering induces magnetoconductance oscillations with flux period $h/2e$.

1. INTERFERENCE IN DISORDERED METAL FILMS

The absorption or emission line spectrum for a gas of atoms provides a unique signature of the quantized electronic orbitals. As already indicated in 1924 by De Broglie, free electrons which are no longer moving into these quantized orbitals, should still retain their quantum–mechanical wave character with the wavelength λ varying inversely proportional to the electron momentum. This was confirmed experimentally in 1927 by Davisson and Germer who showed that, after penetration through a periodic array of atoms in a crystal, the diffraction patterns for an electron beam and an x–ray beam look identical.

In this chapter, we will try to convince the reader that the wave properties can also be observed experimentally for the conduction electrons in a metal. Neglecting band structure and correlation effects the freely moving electrons can be described as plane waves $\Psi(r) \propto exp(ik \cdot r)$ with $|k| = 2\pi/\lambda$ the wave vector. We will restrict ourselves to the low temperature limit where the contribution of the inelastic scattering (at other electrons or phonons) to the electrical conductivity can be neglected. The scattering at lattice defects and impurities will cause an elastic diffusion of the charge carriers. Due to the Pauli principle, only the electrons near the Fermi level contribute to the conductivity which is given by the Einstein relation

$$\sigma_0 = e^2 N(E_F) D. \tag{1}$$

$N(E_F)$ represents the electronic density of states near the Fermi level and $D = \frac{1}{3} v l_{el}$ is the diffusion constant with v the intrinsic electron velocity. The diffusion approach will be valid only for electronic transport on a length scale L much larger than the elastic mean free path l_{el} (*diffusive regime*). On a length scale $L << l_{el}$, we

W. van Haeringen and D. Lenstra (eds.), Analogies in Optics and Micro Electronics, 165–184.
© *1990 Kluwer Academic Publishers. Printed in the Netherlands.*

enter the *ballistic regime*, which is discussed in detail in the chapter by van Houten and Beenakker (Chap.13 in this book).

Eq. (1) results from a statistical interpretation of the conduction process, where the wave propagation of individual electrons is not considered. With the development of the scaling theory for the metal–insulator (Anderson) transition [1], it has become clear that close to the transition, the interference between frequently scattered electron waves causes an important correction of the conductivity. In Fig. 1, we have sketched diffusive electron paths which may influence the four–terminal conductance $G = I_0/(V_+-V_-)$ of a disordered metal film with length L, thickness t, and width w. The current I_0 flows between two large reservoirs with chemical potential μ_L and μ_R. The black dots in Fig. 1 represent the elastic scattering centers where the k vector of the electron changes its direction, giving rise to a random phase shift of the wave function. When only the diffusion of single electron waves (e.g. between points A en B) is considered, the Einstein relation (1) is recovered. On the other hand, one may also take into account the splitting at point C of an electron wave into two partial waves which interfere again at point D. The quantum–mechanical probability P_D that the electron arrives at point D, is given by the square of the sum of the two scattering amplitudes $\psi_u(D)$ and $\psi_l(D)$ which describe the electron in the upper and lower path

$$P_D = |\psi_u|^2 + |\psi_l|^2 + \psi_u^*\psi_l e^{i\theta} + \psi_l^*\psi_u e^{-i\theta} = 2|\psi_u|^2(1+\cos\theta). \qquad (2)$$

We have assumed that the probabilities to diffuse along the upper and the lower path are equal: $|\psi_u|^2 = |\psi_l|^2$. The wave character of the conduction electrons causes extra interference terms whose magnitude is determined by the phase factor θ which depends upon the phase shifts occuring during each of the elastic scattering events. In a *macroscopic* piece of metal with size $L >> l_{el}$, many paths with a different θ value have to be considered, implying that the *ensemble–average* of the interference terms can be neglected and the result (1) remains unaffected.

1.1 Interference in the Backscattering Direction: Weak Electron Localization

The *self–averaging* of the interference effects will not occur for the special class of scattering processes where the electron returns to its original position O (see Fig.1). Since the interference occurs between time–reversed electron states, the phase factor $\theta \equiv 0$ for this *coherent backscattering* [2] and

$$P_0 = 2|\psi_{cw}|^2 + 2|\psi_{cc}|^2 = 4|\psi_{cw}|^2 = 2P_{cl} . \qquad (3)$$

$|\psi_{cw}|^2 (= |\psi_{cc}|^2)$ is the probability for backscattering in the clockwise (counter–clockwise) direction. The quantum–mechanical probability P_0 for backscattering is therefore twice as large as the classical probability P_{cl} for backscattering. Within the diffusion approach, only interference processes on a scale $L_{in}(T) = (D\tau_{in}(T))^{1/2} >> l_{el}$ have to be taken into account, where τ_{in} is the inelastic scattering time. The inelastic scattering changes the electron energy and therefore destroys the phase coherence of the interfering electron waves. The presence of spin–flip scattering (characteristic scattering time τ_{sf}) at magnetic impurities will cause an additional reduction of the phase coherence [3]. $L_{in}(T)$ should therefore be replaced by a more general *phase coherence length* $L_\phi = (D\tau_\phi(T))^{1/2}$. The scattering rate $(\tau_\phi)^{-1}$ will be an appropriate combination of

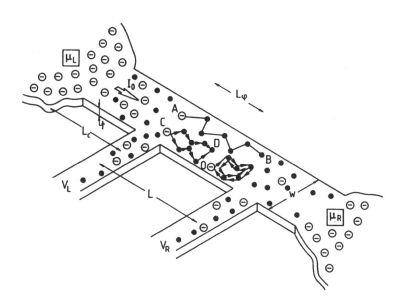

Fig.1   Four–terminal pattern for measuring the conductance of a mesoscopic, disordered metal sample. The conductance is influenced by the classical diffusion (between points A and B), the direct interference between splitted electron waves (between points C and D), and the coherent backscattering (around point O).

the inelastic and spin–flip scattering rates. For typical metals with $l_{el} \sim 10$ nm and containing only a few magnetic impurities (less than 1 ppm), $L_\phi \sim 1$ μm below $1K$.

Due to the rapid decrease of the inelastic scattering at low temperatures ($\tau_{in}(T) \propto T^{-p}$ with the value of p depending on the dominating inelastic scattering mechanism), more backscattering loops will contribute and cause a temperature dependent correction $\Delta\sigma$ of the conductivity σ_0 given by (1). A detailed first–order perturbation calculation using impurity–averaged Green's functions, shows that this conductivity correction strongly depends upon the sample *dimensionality d* [1]

$$\sigma = \sigma_0 + \Delta\sigma = \sigma_0 - C G_{un}[L_\phi(T)]^{2-d} = \sigma_0 - C (e^2/h)[L_\phi(T)]^{2-d}. \quad (4)$$

In order to calculate the exact value of the constant $C \sim 1$, one has to take into account the disorder enhanced Coulomb repulsion which reinforces the localization effect [1] as well as the spin–orbit scattering (characteristic time τ_{so}) which influences the magnitude and the sign of the interference terms [2]. In sharp contrast to the spin–flip scattering, the spin–orbit scattering which is caused by the elastic disorder scattering, does however not destroy the phase coherence! For the quasi two–dimensional case ($d = 2$ and σ coincides with the conductance G of a square film) which occurs when the film thickness t is much smaller than L_ϕ, the correction $\Delta\sigma$ in (4) has to be multiplied by the logarithmic correction factor $\ln[L_\phi(T)/l_{el}]$. This implies that σ will show a logarithmic temperature dependence in thin films. This has been confirmed experimentally [2] for many metals and alloys as well as for the two–dimensional electron gas in Si–MOSFETs. When both the film thickness t and the film width w (see Fig. 1) become much smaller than L_ϕ, the one–dimensional limit ($d = 1$ and σ corresponds to the sample conductance G multiplied by the sample length L) is reached and a stronger power–law temperature dependence of the conductivity is observed [1].

The magic *universal conductance* $G_{un} = e^2/h \simeq (26 \text{ k}\Omega)^{-1}$ appearing in (4), also appears in all of the available quantitative results for interference effects in disordered materials. G_{un} was first introduced by Thouless [4] when he noticed that the Einstein relation (1) implies that the conductance on a scale L can be written as $G(L) = G_{un} V(L)/W(L)$. V is the life–time broadening of the electron states due to the diffusion within the sample, i.e. $V(L) = \hbar/(DL^2)$. W corresponds to the separation between the free electron levels within the volume L^d: $W(L) = [N(E_F)L^d]^{-1}$. The insulating phase with exponentially localized wave functions, will emerge when $V/W \sim 1$, i.e. when $G \sim G_{un}$ and the interference correction $\Delta \sigma > \sigma_0$. The perturbation result (4) is valid only in the *Weak Electron Localization* (WEL) regime [2], where the conductance of the metallic sample is considerably larger than G_{un}. The WEL regime can alternatively also be defined as the regime where $k_F l_{el} >> 1$, i.e. the electronic wavelength λ_F at the Fermi level has to be small when compared to the elastic mean free path l_{el}.

1.2 Direct Interference Between Splitted Electron Waves: The Mesoscopic Regime

While the coherent backscattering is observable in a macroscopic sample, the random phase shift θ appearing in (2), prevents the observation of the *direct interference* between splitted electron waves (e.g. between points C and D in Fig.1). In extremely small samples with size $L \sim l_{el} << L_\phi(T) \sim 1\mu m$, one intuitively expects to observe important conductance changes when the position of the individual impurities is changed. Statistical considerations would suggest that the root–mean–square (rms) amplitude $\Delta \sigma$ of the conductivity correction should decrease when the total number of possible diffusive paths increases. Detailed calculations [5,6] give the surprising result that the *stochastic* self–averaging of the direct interference effects does not occur when increasing the sample size L between l_{el} and L_ϕ. The regime where $l_{el} < L < L_\phi$ is now generally known as the *mesoscopic* regime. For mesoscopic samples, the mean–square deviation around the average conductance G_{av} is independent of the sample size and its dimensionality

$$(\Delta G)^2 = <\delta G^2> = <(G - G_{av})^2> \sim G_{un}^2 = (e^2/h)^2, \qquad (5)$$

where the brackets $<...>$ refer to an average taken over many mesoscopic samples. If one would succeed in preparing mesoscopic samples which only differ by their impurity configuration, (5) predicts that the low–temperature conductance G will fluctuate around G_{av} with rms amplitude $\Delta G \sim e^2/h$. Although it has become rather easy to prepare mesoscopic samples using the existing lithographic techniques, it is impossible to prepare identically shaped samples of the same material with a different impurity configuration. The sample inhomogeneities and imperfections will cause fluctuations which strongly exceed the fluctuation amplitude predicted by (5). Fortunately, the redistribution of the impurities can also be simulated by applying different magnetic fields to a single mesoscopic sample [7]. As we will discuss in detail in the next section, the Aharonov–Bohm effect will allow to continuously tune the phase difference between the interfering electron waves.

2. THE SOLID–STATE AHARONOV–BOHM EFFECT

When a magnetic field B is applied, a classical approach indicates that the Lorentz force $F_L = -ev \times B$, will bend the electron path. This bending is most easily observed via the low–field Hall effect. Within the quantum–mechanical approach, the Lorentz force induces the formation of Landau orbitals. This quantization of the energy levels in a magnetic field, will influence the electronic properties only at low temperatures and/or high magnetic fields where the separation $\hbar\omega_c = e\hbar B/m$ between the levels largely exceeds the thermal energy $k_B T$. On the other hand, the quantum behavior only emerges provided the level separation $\hbar\omega_c$ is also considerably larger than the life–time broadening $\hbar/\tau_{el}$ due to the elastic scattering. In the quantum regime, the famous Shubnikow – De Haas – Van Alphen magnetoresistance oscillations will reflect the sharp structure of the Fermi surface. For the two–dimensional electron gas, the exact quantization of the Hall effect will give an additional confirmation of the quantum behavior. As indicated earlier, we will concentrate here on disordered materials, where the elastic mean free path $l_{el} \sim 10$ nm. In these materials, a magnetic field $B > 100\ T$ would be needed to reach the quantum regime!

2.1 Tuning of the Interference by a Magnetic Field

As first pointed out by Aharonov and Bohm (AB) [8], the presence of a small magnetic field B will also cause a phase shift of the electronic wave function

$$\phi = (e/\hbar) \int A \cdot dl . \tag{6}$$

The phase shift ϕ is simply given by the line integral of the vector potential A along the electron path. The AB effect is a direct consequency of the wave character of the electrons and does not have a classical analogue. Eq. (6) has the remarkable implication that the wave function senses the magnetic field through the vector potential A even in regions where $B = \nabla \times A = 0$. The AB effect is related to the fact that the phase of the wave function, which is calculated from the Schrödinger equation, depends upon the specific choice for the vector potential and is not gauge invariant. On the other hand, the physically observable quantities only depend on the field derived from the potential and are gauge invariant.

The ideal experiment to test the AB tuning of the electronic wave function is presented in Fig.2. This experiment is the solid–state analogue of the well–known two–slit interference experiment. In the absence of a magnetic field, the interference can be observed by having a different path length for the two interfering electron beams. For disordered metal paths, the path length can also be changed by modifying the impurity configuration in the arms of the loop (modify the position of the black dots in Fig.2). Such a modification of the loop geometry or the impurity configuration can never be controlled experimentally. As indicated by (6), the AB effect provides a unique tool to continuously tune the phase difference between the interfering electron waves. When applying (6) between points L and R in Fig. 2, we obtain the extra phase difference ϕ caused by the magnetic field B:

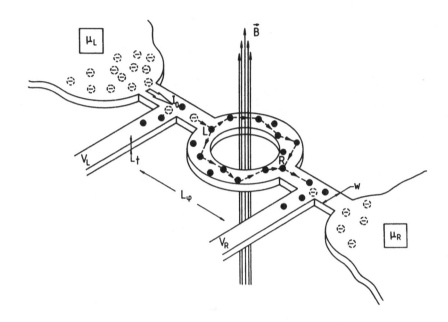

Fig.2   Four–terminal measurement of the Aharonov–Bohm
           magnetoconductance oscillations in a mesoscopic disordered metal loop.
           The interference at point R of the electron waves traveling along both
           arms of the loop, is tuned by the magnetic flux which is completely
           confined to the interior of the loop.

$$\phi(B) = \frac{e}{\hbar} \oint A \cdot dl = \frac{e}{\hbar} \iint B \cdot dS = 2\pi \frac{\Phi_B}{\Phi_0}, \tag{7}$$

where the flux quantum $\Phi_0 = h/e$. The phase difference $\Delta\phi$ varies proportional to
the flux Φ_B enclosed by the loop and is as expected a gauge–invariant quantity
since it is directly related to the transmission probability of the loop. In sharp
contrast to the quantum effects caused by the formation of the Landau levels, the
AB effect also occurs at very small magnetic fields. For a mesoscopic loop with a
diameter of 1 μm, a flux quantum Φ_0 corresponds to a field variation of only 0.05T.

   From relations (2) and (7) we calculate that for the interference process shown
in Fig.2, the probability P_R to arrive at point R is an oscillating function of the
enclosed flux Φ_B:

$$P_R(\Phi_B) = 2|\psi_u|^2[1+\cos(\phi+\theta)] = 2|\psi_u|^2[1+\cos(2\pi\,\Phi_B/\Phi_0 + \theta)]. \tag{8}$$

We want to stress again that this AB modulation with flux period h/e of the
transmission probability occurs even when the magnetic flux is completely confined

to the interior of the loop and the electrons do not experience the Lorentz force F_L. With the development of the transmission electron microscope, it also became possible to perform the AB experiment in vacuum. Under ideal experimental conditions, the electron waves are not scattered and $\theta = 0$. When the electron beam is splitted into two parts using an electrostatic biprism, interference fringes will be formed near the point where the two beams converge. Holographic images of an object can also be taken when this object is placed in one of the interfering beams. In the presence of an homogeneous magnetic field, two effects are observed. First of all, the envelope of the interference pattern is shifted because of the Lorentz force which modifies the path followed by the electrons. At the same time, the position of the interference fringes shifts relative to the envelope of the pattern. Careful experiments [9] have confirmed that the latter AB shift of the fringes with flux period h/e occurs even when the magnetic flux is completely confined to the interior of the electron path.

The AB effect can also be checked directly when the metal loop in Fig.2 becomes superconducting. All the electrons can be described by one single wave function whose phase coherence is no longer affected by the electron scattering and can extend over macroscopic distances. Relations (6) and (7) remain valid provided the electron charge e is replaced by the charge $2e$ of a Cooper pair. Since the superconducting wave function should be single–valued, (7) directly implies the quantization of the magnetic flux which can be trapped inside a hollow superconducting cylinder. The enclosed flux will be an integer multiple of the superconducting flux quantum $\Phi_0/2 = h/2e$. Relations (6) and (7) also allow a straightforward understanding of Josephson and SQUID devices in the presence of a magnetic field.

2.2 Aharonov–Bohm Conductance Oscillations in a Mesoscopic Metal Loop

For the interference experiments in an electron microscope or in a superconductor, the loop formed by the interfering electron beams is sufficiently large to fit an isolated magnetic flux into the loop geometry. For the solid–state analogue of the AB effect we need however a mesoscopic loop with a circonference comparable to $L_\phi \sim 1~\mu m$. This implies that the magnetic field will always penetrate the arms of the loop. The experimental search for the AB oscillations in metal loops already started a few years before the absence of self–averaging in the mesoscopic regime (see (5)) was proven rigorously. The first experiments were sparked by the theoretical work of Gefen, Imry, and Azbel [10] on the possible existence of a solid–state AB effect. These qualitative theoretical predictions were based on the pioneering work of Landauer [11]. In the Landauer model for the electronic conduction, the disorder is treated as a quantum–mechanical tunneling barrier for the conduction electrons. The tunneling model predicts [12] that in an isolated, disordered metal ring (no leads attached) a persistent current, comparable to a superconducting current, can be induced by applying a time–dependent magnetic flux $\Phi_B(t)$, provided there is no phase–breaking scattering present. The classical, resistive behavior only appears when the electrons can loose their phase memory in the electrical leads which are connected to the loop. At the same time, the amplitude of the AB interference effects is largely damped. This strong influence of the measuring leads on the experimentally observed conductance oscillations will be discussed in detail in Section 4.

The first experiments [13] failed to show periodic h/e oscillations. Instead, aperiodic (but reproducible) magnetoconductance fluctuations appeared with an

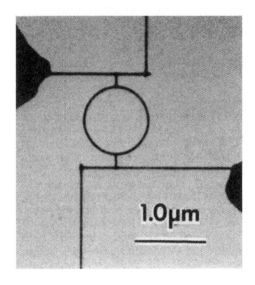

Fig.3   Experimental observation of the solid–state Aharonov–Bohm effect for
             the mesoscopic Au loop shown in the electron micrograph (a). Curve (b)
             is a typical magnetoresistance trace at $T = 50$ mK, showing the periodic
             h/e oscillations and the slower aperiodic fluctuations which are caused
             by the penetration of the magnetic field into the arms of the loop. Curve
             (c) shows the Fourier spectrum of the magnetoresistance trace (b) (from
             Ref. [15]).

amplitude which strongly increased at lower temperatures. As pointed out by
Stone [14], the aperiodic oscillations are caused by the magnetic flux penetrating
the arms of the loop. Direct interference processes occuring within the arms of the
loop will produce sample specific oscillations (*magnetofingerprints*) with different
flux periods depending on the enclosed area. For a loop with a poor aspect ratio,
i.e., w is not much smaller than the loop radius r, the flux Φ_B through the metal
lines forming the loop is comparable to the flux enclosed by the partial waves
which produce the h/e oscillations. The h/e oscillations will therefore be masked
by the aperiodic background, since both effects occur on a comparable
characteristic field scale.

    In 1984, Webb *et al.* [15] succeeded in fabricating a Au loop having a very good
aspect ratio. This loop with radius $r \simeq 0.4$ μm and $w \simeq t \simeq 0.04$ μm was obtained by
contamination lithography in a transmission electron microscope. Fig. 3 shows an
electron micrograph of the mesoscopic loop together with a typical
magnetoresistance trace at $T = 50$ mK, clearly showing the h/e oscillations with
the expected field period of about 0.08 T. Because of the good aspect ratio, the
periodic h/e oscillations can easily be distinguished from the aperiodic fluctuations

which appear as a slowly varying background. Fig. 3 also shows the Fourier transform of the magnetoresistance trace with sharp peaks corresponding to the h/e effect and the second harmonic $h/2e$ effect. These peaks are well separated from the peak near the origin which is caused by the long–wavelength aperiodic fluctuations. The position of the long wavelength peak is consistent with the fact that the aperiodic fluctuations appear as soon as a flux quantum Φ_0 is contained within a phase coherent area wL_ϕ [14]. For the loop shown in Fig.3 with $L_\phi \simeq 1.0$ μm, the minimum wavelength should be of the order of 0.1 T, in agreement with the experimental results.

Alternatively, the minimum wavelength for the aperiodic fluctuations can also be linked to a *correlation field* $B_c \sim \Phi_0/(wL_\phi)$[7]. Physically, B_c corresponds to the field variation which is needed to produce uncorrelated interference patterns within a phase coherent area. Changing the applied magnetic field by an amount B_c has therefore the same effect as modifying the impurity ensemble (*ergodic hypothesis*), B_c can be obtained directly from the half–width of the correlation function $F(\delta B)$ $= \;<G(B)G(B+\delta B)>$, where $<...>$ corresponds to an average over many magnetic fields. Provided the ergodic hypothesis is valid, this average will be identical to an ensemble average.

Although the loop experiment confirms the importance of the direct interference between partial conduction electron waves in the mesoscopic regime, the presence of a pure AB effect can not be proven unambiguously. Since the magnetic flux is also piercing the arms of the loop, the B field interacts with the electrons via the Lorentz force. The clear separation between the h/e signal and the aperiodic background in the Fourier transform does however strongly suggest that the electrons experience the presence of the magnetic flux in the interior of the loop via the Aharonov–Bohm effect. The bending of the electron paths by the Lorentz force is randomized by the diffuse elastic scattering at the film edges. This argument remains valid as long as the cyclotron frequency ω_c is considerably smaller than the elastic scattering rate $(\tau_{el})^{-1}$, i.e. as long as we can neglect the presence of the Landau orbitals.

2.3 Universal Conduction Fluctuations (UCF)

A closer inspection of Fig.3 reveals that the h/e conductance fluctuations have an rms amplitude $\Delta G \sim G_{un} = e^2/h$. Although the Au loop contains more than 10^8 electrons, the impurity averaging is not able to destroy the AB oscillations predicted by (8). The theoretical answer [7] to this puzzling result was already announced in Section 1.2: the intuitive, stochastic self–averaging of the direct interference between splitted electron waves does not occur in the mesoscopic regime. When the magnetic flux is completely confined to the interior of the metal loop, the conductance oscillations around the average value G_0 calculated from the Einstein relation (1), are given by [16]

$$G = G_0 + A_1 G_{un}\cos(2\pi\,\Phi_B/\Phi_0 + \alpha_1)\exp(-\pi r/L_\phi)$$

$$+\; A_2 G_{un}\cos(2\pi\,\Phi_B/(\Phi_0/2) + \alpha_2)\exp(-2\pi r/L_\phi) + \;.... \qquad (9)$$

In (9), r corresponds to the radius of the loop. A_1 and A_2 are sample–dependent correction factors of order unity, whose exact value depends on the strength of the spin–orbit and the spin–flip scattering. α_1 and α_2 are sample–dependent phase

factors and will be discussed in more detail in Section 4. While the second term in
(9) corresponds to the direct interference illustrated in Fig.2, the higher order
terms correspond to electron waves that travel several times around the loop
before interfering. The exponential damping of the magnetoconductance
oscillations which occurs when L_ϕ is smaller than the circonference of the loop,
explains why in Fig.3 the amplitude of the higher harmonic $h/2e$ oscillations is
small when compared to the amplitude of the h/e oscillations.

The magnetic field piercing the arms of the loop, causes aperiodic fluctuations
of the constants A_1 and A_2 with an rms amplitude which is of order unity. When
the ring geometry is replaced by the simple four–terminal geometry shown in
Fig.1, only the aperiodic fluctuations are present. Experimentally, the presence of
such aperiodic fluctuations has been confirmed for mesoscopic samples defined in
thin metal films as well as in the two–dimensional electron gas of semiconductor
devices. Fig.4 shows a typical result at $T = 4.2$ K which has been obtained by
Skocpol et al. [17] for the conductance of a narrow channel ($w \simeq 0.1$ μm) in the
inversion layer of a Si–MOSFET. This experimental result confirms that the
fluctuations for this device with length $L = 0.3$ μm $\simeq L_\phi = 0.25$ μm have the
expected rms amplitude $\Delta G = 0.65$ $e^2/h \sim G_{un}$. The correlation field $B_c = 0.48$ T
calculated from the magnetoconductance trace is in good agreement with the crude
estimate $B_c \sim \Phi_0/(wL_\phi) = 0.16$T.

Fig.4 also shows that the fluctuation pattern strongly depends upon the gate
voltage V_G. It is well known that in MOSFETs, the carrier density of the
two–dimensional electron gas changes with the applied gate voltage. This implies
that also the electron energy E changes, resulting in strongly different interference
patterns. In analogy with the definition of B_c, one can define the correlation energy
E_c as the energy change needed to produce uncorrelated interference patterns.
Obviously, the patterns become totally uncorrelated when $E_c \tau_\phi \sim h$, where τ_ϕ is
the time it takes for the electron to diffuse through the mesoscopic sample. The
correlation energy E_c will then be given by [16]

$$E_c \sim h/\tau_\phi \sim hD/(L_\phi)^2. \tag{10}$$

E_c corresponds to the life–time broadening V of the energy states in a sample with
size L_ϕ (see Section 1.1). From Fig.4, we estimate that the experimental
fluctuation patterns become uncorrelated for $\delta V_g \geq 0.2V$. For the MOSFET device
this corresponds to $E_c = 1.16$ meV $= 4.5k_B T$, in order of magnitude agreement
with the value predicted by (10).

2.4 Stochastic Energy Averaging and Self–Averaging of the Mesoscopic
Fluctuations

Up to now, we have only discussed experimental results which have been obtained
at low temperatures where $k_B T < E_c$. As soon as $k_B T > E_c$, the thermal smearing
of the electrons near the Fermi level will introduce an additional *energy averaging*
of the interference effects. The finite temperature will cause the appearance of
$k_B T/E_c$ uncorrelated interference patterns within the mesoscopic sample. Due to
the stochastic averaging of these patterns, the amplitude of the AB effect
(constants A_1 and A_2 in (9)) will be reduced by a factor $(k_B T/E_c)^{1/2}$ [7].
Experimentally, it has been verified that the AB amplitude decays $\propto T^{-1/2}$ for both
the periodic and the aperiodic oscillations [18].

It is important to note that the energy averaging at finite temperatures occurs

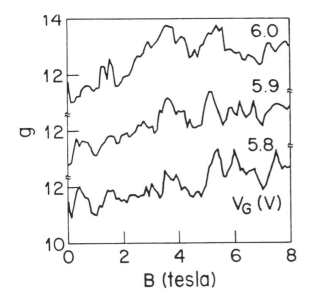

Fig.4   Magnetic field dependence at $T = 4.2$ K of the normalized conductance
             $g = G/G_{\mathrm{un}}$ at three gate voltages for a submicron channel defined in a
             Si inversion layer (from Ref. [17]).

independently of the self–averaging of the interference effects in samples with a
size $L > L_\phi$. For larger samples, we intuitively expect that the self–averaging will
cause a reduction of the oscillation amplitude ΔG by a factor $N^{1/2}$, where N is the
number of mesoscopic units within the sample. More detailed calculations indicate
that the details of the averaging process depend upon the sample dimensionality
[7,19].
        The self–averaging of the interference effects in the one–dimensional limit
$(w,t << L_\phi)$, can be tested experimentally by fabricating an array of N mesoscopic
metal loops in series [20], where the interference effects occuring for the individual
loops are uncorrelated and will add up stochastically. Alternatively, the
self–averaging can also be tested by studying the amplitude ΔG of the aperiodic
magnetoconductance oscillations in fine line samples with a length $L > L_\phi$. As well
for narrow MOSFET channels as for dirty metal samples, the stochastic averaging
process is confirmed [21]. Since the average conductance G_0 of the sample decreases
as $N^{-1} = L_\phi/L$ while the aperiodic oscillations of the individual segments with
length L_ϕ add up incoherently, i.e. $\propto N^{-1/2}$, the fluctuation amplitude for the total
sample will be $\Delta G \sim (e^2/h)(L_\phi/L)^{3/2} \propto N^{-3/2}$.

3. INTERFERENCE PROCESSES IN THE MACROSCOPIC REGIME

As mentioned in Section 1.1, interference processes in the backscattering direction
survive even on a macroscopic scale and cause the WEL correction to the
conductivity. For the loop geometry shown in Fig.2, these interference processes
correspond to two partial waves which travel completely around the loop and
interfere again at point L. In (9), the flux quantum $\Phi_0 = h/e$ has to be replaced by

the superconducting flux quantum $\Phi_0/2 = h/2e$. Consequently, the magnetoconductance oscillations resulting from the WEL correction will have the fundamental flux period $h/2e$.

In a thin metal film, where the loop structure is not present, many different areas will be enclosed by the diffusing electron waves. The conductance oscillations are therefore washed out and are replaced by a uniform background magnetoconductance. This magnetoconductance will become observable as soon as the superconducting flux quantum $h/2e$ fits into an area $(L_\phi)^2$. Since $L_\phi \sim 1\ \mu m$ at low temperatures, the anomalous WEL magnetoconductance will become important for fields $B > 0.01$ T. In the absence of spin–orbit scattering, the destruction of the constructive interference by the magnetic field will induce a positive magnetoconductance (negative magnetoresistance). For strong spin–orbit scattering ($\tau_{so} << \tau_\phi$), the interference in the backscattering direction is on the average destructive, and the magnetoconductance becomes negative (positive magnetoresistance). Numerous experiments [2] on disordered thin metal films have shown that an analysis of the low–field magnetoconductance in a perpendicular field is a unique and extremely powerful tool to study characteristic scattering processes: the inelastic scattering due to the electron–electron or the electron–phonon interaction, the spin–flip scattering caused by the Kondo effect in dilute magnetic alloys, the enhancement of the spin–orbit scattering at a disordered film surface, etc... .

As first noticed in 1981 by Altshuler, Aronov and Spivak [22], the $h/2e$ oscillations will also survive for a thin–walled metal cylinder with diameter $\phi \sim L_\phi$ and length $H \sim 1$ cm (see Fig. 5a). When such a cylinder is placed in an axial magnetic field, all the electron waves that start e.g. at point P and are scattered elastically around the cylinder, will enclose the same magnetic flux $\Phi_B = \pi \phi^2 B/4$ and contribute to the $h/2e$ oscillations. On the other hand, electron waves starting e.g. at point P$'$ which do not travel completely around the cylinder, enclose zero magnetic flux and do not contribute to the magnetoconductance. While the cylinder diameter has a mesoscopic size, the cylinder length H is still macroscopic: $H >> L_\phi$. Altshuler et al. applied the Green's function approach which had been used for the derivation of (4), to calculate the magnetoconductance oscillations for the cylinder geometry. When the conductance is measured between the top and the bottom of the cylinder, the amplitude $\Delta \sigma$ of the AAS $h/2e$ oscillations is given by [22]

$$\Delta \sigma \simeq -(\Delta R_\square/R_\square^2) \simeq - G_{un} K_0(2\pi r/L_\phi). \tag{11}$$

$R_\square$ is the resistance per square of the metal cylinder wall and $K_0(x)$ is the McDonald function. When the cylinder circonference $\pi\phi$ becomes larger than the phase coherence length L_ϕ, the McDonald function results in an exponential damping of the oscillations. Apparently, the amplitude of the $h/2e$ resistance oscillations can be enhanced considerably by increasing the disorder so that the resistance per square $R_\square$ becomes comparable to $(G_{un})^{-1}$. Unfortunately, the inelastic diffusion length L_{in} varies inversely proportional to $R_\square$ in disordered metal films [2]. The exponential damping caused by the decrease of the phase coherence length L_ϕ will be much stronger than the enhancement of the oscillation amplitude caused by a larger $R_\square$ value. It is also important to note that the AAS $h/2e$ oscillations are a particular example of the higher harmonic $h/2e$ Aharonov–Bohm oscillations (third term in (9)), which also include interference between electron states that do not obey the time–reversal symmetry.

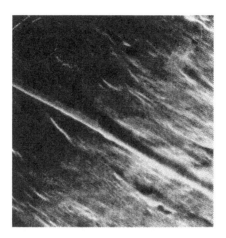

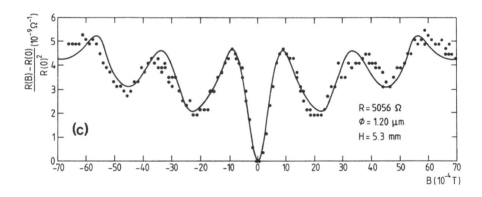

Fig.5　　　Experimental observation of the magnetoconductance oscillations with
flux period $h/2e$ for a thin–walled Mg cylinder: (a) shows the schematics
of the measuring geometry, (b) is an electron micrograph (magnification
× 2200) showing the stretched MG cylinder with diameter $\phi = 1.20\ \mu m$
near one of the electrical contacts, and (c) shows the normalized
low–field magnetoresistance at $T = 1.5$ K (from Ref. [23]).

Experimentally, thin–walled metal cylinders with diameter $\phi \sim 1$ μm can be fabricated by evaporating a 20 nm thick metal film onto a quartz fiber which is stretched over a hole (diameter ~ 1 cm) in a glass substrate. During the metal evaporation, the substrate is rotated in a reduced helium atmosphere to ensure a uniform thickness of the cylinder wall. Fig. 5b shows an electron micrograph of a Mg cylinder with diameter $\phi = 1.20$ μm at a magnification × 2200. Fig. 5c represents the low–field magnetoresistance trace (field parallel to the fiber axis) for this Mg cylinder at $T = 1.5$K. The experimental results, obtained by Gijs, Van Haesendonck, and Bruynseraede in 1984 [23], clearly confirm the existence of the $h/2e$ oscillations which were observed for the first time by Sharvin and Sharvin in 1981 [24]. As shown by the full curve in Fig.5c, the $h/2e$ oscillations can be nicely fitted by the AAS theory when we assume a phase coherence length $L_\phi \simeq 2.0$ μm.

The reduction of the $h/2e$ oscillation amplitude at higher magnetic fields is mainly due to the magnetic field penetrating the cylinder walls. The result (11) is valid only for cylinders with wall thickness $t \rightarrow 0$. For a finite wall thickness, a high enough magnetic field causes different phase shifts for electron waves traveling around the cylinder along different paths. The $h/2e$ oscillations will be destroyed as soon as a superconducting flux quantum $h/2e$ fits into the area $\pi\phi t$ (B should not exceed 0.03 T for the cylinder shown in Fig.5). The strong damping of the $h/2e$ oscillations at higher fields becomes more obvious in Fig. 6 where also the results obtained at higher temperatures are shown. The uniform WEL background magnetoresistance is largely caused by the fact that the magnetic field can not be perfectly aligned with the cylinder axis. This implies that the backscattering loop starting at point P$'$ in Fig.5a will start to contribute to the low–field background magnetoresistance. At this point, we also note that the magnetoresistance for a long cylinder and a large periodic array of mesoscopic loops [25] look similar. This can be understood when we imagine that the cylinder can be chopped into a very large number of mesoscopic loops. Due to the important self–averaging, the sample specific h/e oscillations can no longer be resolved.

The results shown in Fig.6, also clearly illustrate the exponential damping of the AAS oscillations at higher temperatures. At higher temperatures, the phase coherence length for electron waves traveling around the cylinder, rapidly decreases due to the inelastic electron–electron and electron–phonon scattering. This results in an exponential damping of the AAS oscillations above $T = 1.5$ K where $L_\phi(T) \simeq L_{in}(T)$ becomes considerably smaller than the cylinder circonference $\pi\phi$. As shown in the insert of Fig.6, the phase coherence length L_ϕ calculated from the oscillation amplitude using (11) is nearly identical to the L_ϕ values obtained from the magnetoresistance measurements for a flat Mg film. This clearly confirms that the $h/2e$ oscillations appearing for the cylinder geometry and the WEL effects have the same physical origin: the coherent backscattering.

The fact that it took more than two years [23,24] to confirm the original experiments of Sharvin and Sharvin, was largely due to the problem of obtaining thin films with a sufficiently large L_ϕ value at low temperatures. Experimentally, it has become clear that for pure metal films ($R_\square \sim 1$ $\Omega/\square$), L_ϕ will saturate towards a constant low–temperature value which is determined by the presence of residual magnetic impurities [3] or a paramagnetic oxide layer [26]. WEL experiments confirm the validity of the theoretical result

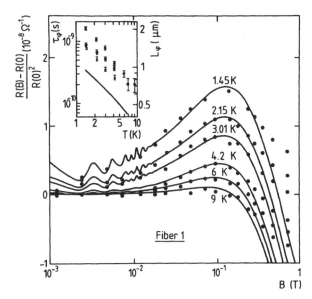

Fig.6   Normalized magnetoresistance at higher fields and at higher
         temperatures for the thin–walled Mg cylinder which is also shown in
         Fig.5. The inset shows the temperature dependence of the phase
         coherence length L_ϕ obtained for different cylinders as well as for a flat
         Mg film (solid line) (from Ref. [23]).

$$(\tau_\phi(T))^{-1} = (\tau_{\rm in}(T))^{-1} + 2(\tau_{\rm sf})^{-1} . \tag{12}$$

The presence of a small fraction of magnetic impurities will cause a finite value
$L_\phi(T \to 0) = (D\tau_{\rm sf})^{1/2} \sim 1\ \mu m$ even for the purest metal film having a clean
surface. Relation (12) may also explain why the first experiments on mesoscopic
Au loops [15] failed to show the AAS $h/2e$ oscillations, while the AB h/e
oscillations were clearly present. As indicated by (9) and (11), the $h/2e$ oscillations
are much more strongly damped by the finite L_ϕ value, since the interference in
the backscattering direction occurs over a length scale which is twice as large as for
the direct interference. Experiments on metal loops with a larger L_ϕ value have
confirmed the coexistence of both the AAS $h/2e$ oscillations and the AB h/e
oscillations at low magnetic fields [20,27]. We should also note that the result (12)
is strictly valid only for the WEL interference. An important fraction of the direct
interference processes in mesoscopic samples occurs however between states that
are not time–reversed and survive in a high magnetic field. This explains why the
h/e oscillations can be observed in magnetic fields as high as 15 T [28]. On the
other hand, theoretical calculations [7,29] indicate that all the direct interference
processes (including the processes without the time–reversal symmetry) are equally
destroyed by the spin–flip scattering at magnetic impurities. The L_ϕ value
obtained from WEL experiments will therefore still provide a good guess for the
coherence length governing the mesoscopic fluctuations at higher magnetic fields.
At very high magnetic fields, the spin–flip scattering process becomes inelastic due
to the Zeeman splitting. The L_ϕ value will then become field dependent and the

result (12) should be corrected.

Since the AAS $h/2e$ oscillations result from the correlation between time–reversed states which is strongly affected by a magnetic field or by the presence of magnetic impurities [3], the AAS effect is very similar to the flux–periodic phenomena in superconducting samples. For a superconducting thin–walled cylinder, the tuning of the superconducting wave function by the magnetic field will also result in resistance oscillations near the superconducting transition. These oscillations were already observed in 1962 by Little and Parks [30] and are present as long as the superconducting coherence length ξ is comparable to the cylinder circumference. Since ξ decreases much faster than the phase–breaking length above the superconducting transition, the Little–Parks $h/2e$ oscillations will be dominated by the AAS $h/2e$ oscillations at higher temperatures [31].

4. NON–LOCALITY OF THE INTERFERENCE EFFECTS

The presence of the AB oscillations in mesoscopic structures clearly demonstrates the wave character of the conduction electrons. Since the interference patterns extend over a length L_ϕ, this inevitably implies the non–local nature of the conductance oscillations [32]. For a *four–terminal measurement*, the interference extends into the voltage and current probes which are also disordered. While the average conductance G_0 only depends upon the properties of the material between the voltage probes, the conductance fluctuatons δG are influenced by all electron paths within a distance L_ϕ of the voltage probes. When the current leads are connected to large contact paths with a high conductance (reservoirs with chemical potential μ_L and μ_R) as shown in Fig. 1 and Fig. 2, the distance L_c separating the voltage probes from the large paths, strongly influences the amplitude of the conductance fluctuations when $L_c < L_\phi$. For $L_c \rightarrow 0$, both the voltage and current probes disappear into the large contact areas and the geometry shown in Fig. 1 and Fig. 2 will correspond to a *two–terminal measurement*, for which non–local effects can be neglected. As expected, also the WEL corrections which appear at low magnetic fields, are strongly influenced by the details of the probe geometry [33].

Umbach et al. [34], have recently provided a very graphic and easy to understand demonstration of the non–local interference effects. As shown by the electron micrograph in Fig.7, two identical mesoscopic Au lines (length 2 μm and linewidth 62 nm) have been prepared by contamination lithography in a transmission electron microscope. A loop with a diameter of 0.7 μm has been connected to one of the lines near the junction of a pair of current and voltage leads. Fig.7 also shows a typical magnetoconductance trace for the two lines. While the single line produces only aperiodic fluctuations which are caused by the penetration of the magnetic field into the line segments, the line with the externally connected loop shows additional h/e oscillations.

The four–terminal geometry is also responsible for the *asymmetry* of the magnetoconductance which is observed in the mesoscopic regime. Apparently, this implies a violation of Onsager's principle requiring $G(B) = G(-B)$. As pointed out by Büttiker [35], the measured voltage will always include an *anti–symmetric* Hall voltage V_H ($V_H(B) = -V_H(-B)$), even when the voltage probes are not on opposite sides of the fine line structure. Since this mesoscopic Hall voltage fluctuates independently of the *symmetric*, longitudinal voltage with an amplitude corresponding to $\Delta G \sim G_{un}$, the apparent violation of Onsager's principle can be explained. In (9), the presence of the mesoscopic Hall voltage was already

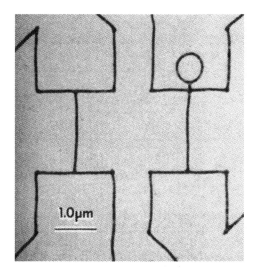

Fig.7   Four–terminal magnetoconductance measurement for a mesoscopic Au
           line with and without a loop attached to the outside of the contact
           probes. While the line without the loop only shows aperiodic
           fluctuations, the line with the loop clearly shows additional periodic
           oscillations with flux period h/e (from Ref. [34]).

anticipated by the presence of the two arbitrary phase factors α_1 and α_2. The
theoretical calculations [16] predict that the asymmetry should be absent for a
two–terminal geometry; α_1 and α_2 are equal to 0 or π.
    In order to determine quantitatively the correction of the fluctuation amplitude
caused by the four–terminal geometry, Benoit et al. [36] have prepared
one–dimensional multiprobe lines of Au and Sb (width $w < 0.1\ \mu$m), where the
sample length L can be varied by choosing different pairs of voltage probes. The
results of the magnetoconductance measurements are summarized in Fig.8, where
the amplitude of the voltage fluctuations has been plotted as a function of the

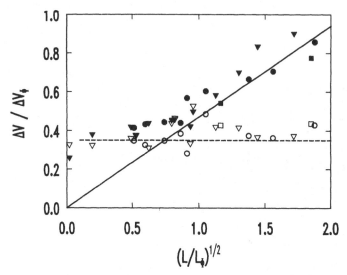

Fig.8   Experimental results for the voltage fluctuation amplitude ΔV
          (normalized by the fluctuation amplitude for a sample with length L_ϕ)
          for one–dimensional Au and Sb lines (width $w < 0.1$ μm), as a function
          of the reduced sample length $(L/L_\phi)^{1/2}$. The closed symbols represent
          the results obtained for the longitudinal, symmetric voltage fluctuations.
          The open symbols are the results for the mesoscopic, anti–symmetric
          Hall voltage (from Ref. [36]).

reduced sample length $(L/L_\phi)^{1/2}$. The rms amplitude ΔV has been normalized by
the rms amplitude ΔV_ϕ corresponding to a sample with length $L = L_\phi$. In this
way, a universal behaviour, independent of the sample properties (material, width
w, L_ϕ value, ...), is obtained. Due to the presence of the mesoscopic Hall effect, the
voltage fluctuations will contain a symmetric as well as an anti–symmetric
component.

    The results for the symmetric component are shown by the closed symbols in
Fig.8. As already indicated in Section 2.3, both theory and experiment are
consistent with the existence of material–independent Universal Conductance
Fluctuations (UCF) with rms amplitude $\Delta G \sim G_{un} = e^2/h$ for $L \sim L_\phi$. The concept
of the UCF (see (5) and (9)) is expected to remain valid for a two–terminal
measurement [7] even when $L << L_\phi$. The results shown in Fig. 8 are however
consistent with ΔV being a constant for $L < L_\phi$. i.e. $\Delta G \propto L^{-2}$, indicating that the
conductance fluctuations are no longer universal for the four–terminal geometry.
Obviously, this non–universality can be easily understood when one considers that
due to the non–local nature of the interference processes, $\Delta V(L)$ will always
correspond to $\Delta V(L = L_\phi)$, while the average voltage $V_0 = I_0/G_0 \propto L$. This
interpretation will fail for very short samples with $L < l_{el} \sim 10$ nm, where the
average conductance can no longer be defined. Due to the limitations of the
lithographic processes, it is not possible to probe this ballistic limit for the
disordered samples which have been discussed in this chapter. We note that the
variation $\Delta V \propto L^{1/2}$ observed for $L > L_\phi$, is consistent with the stochastic
self–averaging which has been discussed in Section 2.4. Similar results have also

been obtained for narrow channels defined in Si inversion layers [37].

The open symbols in Fig.8 refer to the fluctuations of the anti–symmetric mesoscopic Hall voltage V_H. In the mesoscopic regime ($L < L_\phi$), the fluctuation amplitude for V_H is comparable to the longitudinal fluctuation amplitude. For larger samples ($L > L_\phi$), the fluctuation amplitude for V_H remains independent of the sample length L. This implies that the longitudinal, symmetric voltage fluctuations will gradually dominate the Hall–like voltage fluctuations and the apparent asymmetry of the magnetoconductance will disappear for larger samples.

The results presented in Fig.7 and Fig.8 imply that a four–terminal measurement fails in the mesoscopic regime. Consequently, the classical scaling down approach to modeling small circuits will have to be modified to take into account possible interference effects. On the other hand, when the periodic AB oscillations can be made large enough, it is also conceivable that devices based on the quantum interference will become important in the future. Such devices will probably operate in the ballistic regime where the conductance oscillations may become comparable to the average conductance.

Finally, we also note that the sensitivity of the interference to the specific impurity configuration may provide a possible explanation for the presence of $1/f$ noise in metals [6]. The movement of the defects in a metal or the presence of charge traps in a semiconducting device at finite temperatures, will result in discrete conductance jumps comparable to G_{un} for mesoscopic samples. In macroscopic samples, the stochastic averaging can give rise to a $1/f$ noise spectrum. Although this mechanism is unable to explain the observed noise at room temperature, it gives a good description of the excess $1/f$ noise which appears in very dirty materials at low temperatures [38]. This noise is reduced by a factor of 2 when a small magnetic field $B \sim \Phi_0/(L_\phi)^2$ is applied, confirming the destruction of the interference processes between time–reversed states [29].

REFERENCES

[1]   P.A. Lee and T.V. Ramakrishnan, Rev. Mod. Phys. 57 (1985) 287.
[2]   G. Bergmann, Phys. Rep. 107 (1984) 1.
[3]   R.P. Peters, G. Bergmann and R.M. Mueller, Phys. Rev. Lett. 58 (1987) 1964; C. Van Haesendonck, J. Vranken and Y. Bruynseraede, Phys. Rev. Lett. 58 (1987) 1968.
[4]   D.J. Thouless, Phys. Rev. Lett. 39 (1977) 1167.
[5]   B.L. Altshuler, JETP Lett. 41 (1985) 648; B.L. Altshuler and B.Z. Spivak, JETP Lett. 42 (1986) 447.
[6]   S. Feng, P.A. Lee and A.D. Stone, Phys. Rev. Lett. 56 (1986) 1960; *ibid* 56 (1986) 2772(E).
[7]   P.A. Lee and A.D. Stone, Phys. Rev. Lett. 55 (1985) 1622; P.A. Lee, A.D. Stone and H. Fukuyama, Phys. Rev. B35 (1987) 1039.
[8]   Y. Aharonov and D. Bohm, Phys. Rev. 115 (1959) 485.
[9]   A. Tonomura, N. Osakabe, T. Matsuda, T. Kawasaki, J. Endo, S. Yano and H. Yamada, Phys. Rev. Lett. 56 (1986) 792.
[10] Y. Gefen, Y. Imry, and M. Ya. Azbel, Phys. Rev. Lett. 52 (1984) 129.
[11] R. Landauer, IBM J. Res. Develop. 1 (1957) 223.
[12] M. Büttiker, Y. Imry and M. Ya. Azbel, Phys. Rev. A30 (1984) 1982.
[13] C.P. Umbach, S. Washburn, R.B. Laibowitz and R.A. Webb, Phys. Rev. B30 (1984) 4048; G. Blonder, Bull. Am. Phys. Soc. 29 (1984) 535.
[14] A.D. Stone, Phys. Rev. Lett. 54 (1985) 2692.

[15] R.A. Webb, S. Washburn, C.P. Umbach and R.B. Laibowitz, Phys. Rev. Lett. 54 (1985) 2696.

[16] A.D. Stone and Y. Imry, Phys. Rev. Lett. 56 (1986) 189.

[17] W.J. Skocpol, P.M. Mankiewich, R.E. Howard, L.D. Jackel, D.M. Tennant and A.D. Stone, Phys. Rev. Lett. 56 (1986) 2865.

[18] S. Washburn, C.P. Umbach, R.B. Laibowitz and R.A. Webb, Phys. Rev. B32 (1985) 4789.

[19] B.L. Altshuler and D.E. Khmelnitskii, JETP Lett. 42 (1986) 359.

[20] C.P. Umbach, C. Van Haesendonck, R.B. Laibowitz, S. Washburn and R.A. Webb, Phys. Rev. Lett. 56 (1986) 386.

[21] J.C. Licini, D.J. Bishop, M.A. Kastner and J. Melngailis, Phys. Rev. Lett. 55 (1985) 2987; S.B. Kaplan and A. Hartstein, Phys. Rev. Lett. 56 (1986) 2403; V.T. Petrashov, P. Reinders and M. Springford, JETP Lett. 45 (1987) 720.

[22] B.L. Altshuler, A.G. Aronov and B.Z. Spivak, JETP Lett. 33 (1981) 94.

[23] M. Gijs, C. Van Haesendonck and Y. Bruynseraede, Phys. Rev. Lett. 52 (1984) 2069.

[24] D. Yu. Sharvin and Yu. V. Sharvin, JETP Lett. 34 (1981) 272.

[25] B. Pannetier, J. Chaussy, R. Rammal and P. Gandit, Phys. Rev. Lett. 53 (1984) 717; G.J. Dolan, J.D. Licini and D.J. Bishop, Phys. Rev. Lett. 56 (1986) 1493.

[26] J. Vranken, C. Van Haesendonck and Y. Bruynseraede, Phys. Rev. B37 (1988) 8502.

[27] V. Chandrasekhar, M.J. Rooks, S. Wind and D.E. Prober, Phys. Rev. Lett. 55 (1985) 1610.

[28] A.D. Benoit, S. Washburn, C.P. Umbach, R.B. Laibowitz and R.A. Webb, Phys. Rev. Lett. 57 (1986) 1765.

[29] A.D. Stone, Phys. Rev. B 39 (1989) 10736.

[30] W.A. Little and R.D. Parks, Phys. Rev. Lett. 9 (1962) 9.

[31] M. Gijs, C. Van Haesendonck and Y. Bruynseraede, Phys. Rev. B30 (1984) 2964; J.M. Gordon, Phys. Rev. B30 (1984) 6770.

[32] S. Maekawa, Y. Isawa and H. Ebisawa, J. Phys. Soc. Jpn. 56 (1987) 25; M. Büttiker, Phys. Rev. B35 (1987) 4123; H.U. Baranger, A.D. Stone and D.P. DiVincenzo, Phys. Rev. B37 (1988) 6521.

[33] B. Douçot and R. Rammal, Phys. Rev. Lett. 55 (1985) 1148; V. Chandrasekhar, D.E. Prober and P. Santhanam, Phys. Rev. Lett. 61 (1988) 2253; P. Santhanam, Phys. Rev. B39 (1989) 2541.

[34] C.P. Umbach, P. Santhanam, C. Van Haesendonck and R.A. Webb, Appl. Phys. Lett. 50 (1987) 1289.

[35] M. Büttiker, Phys. Rev. Lett. 57 (1986) 1761.

[36] A. Benoit, C.P. Umbach, R.B. Laibowitz and R.A. Webb, Phys. Rev. Lett. 58 (1987) 2343.

[37] W.J. Skocpol, P.M. Mankiewich, R.E. Howard, L.D. Jackel, D.M. Tennant and A.D. Stone, Phys. Rev. Lett. 58 (1987) 2347.

[38] N.O. Birge, B. Golding and W.H. Haemmerle, Phys. Rev. Lett. 62 (1989) 195.

C. Van Haesendonck is with the Laboratorium voor Vaste Stof–Fysika en Magnetisme, Katholieke Universiteit Leuven, B–3030 Leuven, Belgium. He is a Research Associate of the Belgian National Fund for Scientific Research.

PHASE–SENSITIVE VOLTAGE MEASUREMENTS

M. Büttiker

The electro–chemical potential drop across a single scatterer in an otherwise perfect conductor measured by weakly coupled point contacts is investigated. Phase sensitive probes yield a spatially oscillating voltage over distances of a phase–coherence length away from a scatterer. For a one–channel conductor these oscillations are determined by the Fermi wavelength. In a many channel conductor the longest oscillation period is determined by the smallest difference of the longitudinal momenta of differing channels. Averaging over these oscillations yields a potential drop which is smaller than that obtained by assuming a phase–insensitive probe.

1. INTRODUCTION

In conductors so small, that electrons retain phase–memory over distances which are large compared to the distance over which one wishes to determine the voltage drop, the measured voltage is sensitive to the phase of the electrons. Standard resistance measurements are made in a four–terminal set–up: two terminals are used as carrier source and sink and two terminals are used to measure a voltage difference. Below we analyze a simple geometry, shown in Fig.1, to illustrate the nature of phase sensitive voltage measurements. In the four–terminal conductor of Fig.1, carriers incident from terminal 1 or 2 can reach a barrier almost unimpeded by the presence of terminals 3 and 4. Carriers incident from terminal 3 and 4 are reflected with a probability close to one. The interesting quantum mechanical interference effect in this geometry arises from the fact that carriers reflected at the barrier are phase coherent with the incident carriers. Below we focus on the effect of this interference on the resistances which can be measured at the four–terminal conductor shown in Fig.1.

Our discussion is based on the transmission of electrons through the sample and reflection of electrons at the sample. The connection between electrical transport and transmission of carriers through potential hills is almost as old as quantum mechanics itself. Shortly after Oppenheimer's prediction of quantum mechanical tunneling [1], Fowler and Nordheim treated electric fieldemission as a transmission problem of electrons through a potential barrier formed by the work function and the applied electric potential [2]. Strangely, many decades were to elapse before a theory of electric conduction in terms of transmission coefficients would be taken seriously outside the narrow context of quantum tunneling. Frenkel [3] in 1930 treats metal–metal interfaces as a transmission problem and also addressed the non–linear and rectifying properties of such contacts. He also mentions and discusses the application of his theory to granular films. In 1957 Landauer returned to this problem. Landauer's paper focuses on the electrostatic field drop around impurities and potential barriers in the presence of transport [4]. Such local fields are a consequence of accumulation and depletion of charges near the scattering

W. van Haeringen and D. Lenstra (eds.), Analogies in Optics and Micro Electronics, 185–202.

(a)

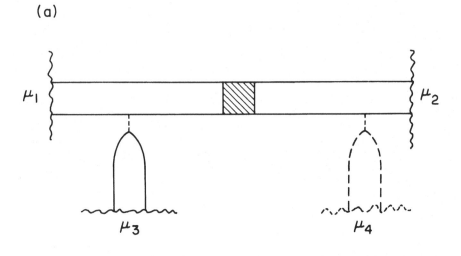

(b)

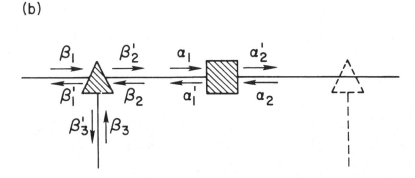

Fig. 1   (a) Multiprobe conductor with one or more weakly coupled probes.
            Aside from the probes, the only scattering process considered is that due
            to a barrier. (b) Schematic representation of the conductor depicted in
            Fig.1a. α and β are the current amplitudes which need to be calculated
            to determine the overall transmission probabilities.

object. Specializing his result for a conductor with only two states at the Fermi
energy, Landauer [5] made the interesting prediction that the resistance of an
obstacle with transmission coefficient T is proportional to $(1-T)/T$. This is an

intriguing result. If the transmission coefficient is 1, there is no obstacle to electrical conduction, and hence the resistance should be zero (the conductance infinite). With his 1970 paper, and related papers by Erdös and Herndon [6], an important step in the theory of electrical conduction based on transmission coefficients was taken: the approach is applied not only to a single obstacle but also to quantum transport through a sequence of scatterers. It is the purpose of this paper to investigate the $(1-T)/T$ result of Landauer from the point of view of a more recent advance alluded to already above: the incorporation of the "measuring device" into the determination of resistances of micro–electronic conductors. The sample with all its terminals is viewed as a coherent entity [7,8]. The geometry of Fig.1 has been analyzed by Engquist and Anderson [9] assuming that at the junctions it is only currents but not their quantum mechanical amplitudes which matters. It is, therefore, of interest to compare a quantum mechanical calculation with the results of Engquist and Anderson.

The long march to formulate electric conductance in terms of a scattering problem is rather perplexing, when compared to a related field: the propagation of electromagnetic waves in wave guides. In this field: the notion that the entire network can be described in terms of transmission and reflection at each incident port, was taken with apparent ease [10]. The close connection between electromagnetic wave propagation and the Schrödinger equation on the other hand might have given theorists concerned with electric conduction at least a hint. There are, however, fundamental differences between photons and electric carriers. Photons are not conserved but charge is. Photons obey Bose–Einstein statistics and electrons obey Fermi–Dirac statistics. It is not enough to ponder the close relationship of Maxwell's equation and the Schrödinger equation. Electrons are coupled to the Maxwell fields via current and charge density. At the very least this coupling gives rise to questions about the screening of piled up electric charges, an issue which does not complicate transmission of electromagnetic waves.

Interestingly, the approach which we take here, does avoid the problem of screening, at least as long as we are concerned with electric conduction in the linear response regime. We assume that the effective equilibrium one–electron potential for the conductor to be considered has been obtained. (In practice this requires the solution of a self–consistent potential problem if the starting point is a bare potential). We assume that the conductor shown in Fig.1 is connected to electron reservoirs at its terminals. These electron reservoirs are not only carrier sources and sinks, but are also a source and sink of energy. They provide for inelastic relaxation and phase–randomization. We assume furthermore, that these terminals are so closely spaced, that transmission of carriers from one reservoir to another can be considered to be elastic. An electron wave incident from one reservoir on the conductor typically gives rise to transmitted waves into all the other terminals and gives rise to a reflected wave back into the reservoir from which the wave is incident. The key point is now that the resistances of this conductor can be expressed in terms of the chemical potentials of the reservoirs and the probabilities at *equilibrium* for transmission from one reservoir to the other. Ref. 7 finds for the current incident on the conductor from reservoir i at the chemical potential μ_i,

$$I_i = \frac{e}{h}[(M_i - R_{ii})\mu_i - \sum_{j \neq i} T_{ij}\mu_j],$$ (1)

where the transport coefficients are the total probability for reflection R_{ii} for a carrier incident in lead i and the total probability for transmission from lead j to lead i, T_{ij}. In (1), M_i is the number of channels in the reservoir i at the connection to the conductor. Let us now consider the situation where lead 1 is the source of carriers and lead 2 is the carrier sink. Lead 3 is used to measure the voltage. The infinite impedance of a voltmeter requires that no net current flows in lead three. Using (1) yields [11],

$$\mu_3 = \frac{T_{31}\mu_1 + T_{32}\mu_2}{T_{31} + T_{32}} . \tag{2}$$

Here we have made use of current conservation: The rows and columns of the matrix of transport coefficients in (1) add up to zero. Eq. (2) expresses a voltage measurement in terms of global transport coefficients. The measured voltage is not a local property: it is transmission from one terminal to another which matters.

Suppose, the voltage probe is to the right of the scatterer (indicated by broken lines in Fig. 1a). Denote the chemical potential measured by the probe in this position by μ_4. It is, similarly to μ_3, determined by a pair of transmission probabilities,

$$\mu_4 = \frac{T_{41}\mu_1 + T_{42}\mu_2}{T_{41} + T_{42}} . \tag{3}$$

To derive (2) and (3), we have each time assumed the presence of one probe only. This is obviously correct, if we imagine a single moveable probe. It is also correct, if we imagine two probes which are weakly coupled to the conductor and take the coupling strength as a small parameter [11]. The two voltage measurements give a voltage drop across the scatterer given by

$$\mu_3 - \mu_4 = \frac{T_{31}T_{42} - T_{32}T_{41}}{(T_{31} + T_{32})(T_{41} + T_{42})} (\mu_1 - \mu_2). \tag{4}$$

If the probes 3 and 4 are only weakly coupled to the conductor, the net current flowing from terminal 1 to terminal 2 is to lowest order in the coupling strength unaffected by the presence of the probes. We can then neglect terms proportional to μ_3 in the equation for I_1 and I_2 and find $I_1 = -I_2 = (e/h)T(\mu_1 - \mu_2)$. Here T is an abbreviation for $T_{21} = T_{12}$. That the two transport coefficients are equal is a consequence of microreversibility for a two–terminal conductor. T characterizes the total transmission of carriers permitted by the scatterer. Thus the measured resistance is given by [11]

$$\mathcal{R} = \frac{(\mu_3 - \mu_4)}{eI} = \frac{h}{e^2} \frac{1}{T} \frac{T_{31}T_{42} - T_{32}T_{41}}{(T_{31} + T_{32})(T_{41} + T_{42})} . \tag{5}$$

Instead of using (2) twice (for the probe to the left and then for the probe to the right of the barrier) we could equally well have started from a four probe conductor with four terminals described by (1) with i = 1,2,3,4. Thus (5) is a simple limit of a four terminal measurement with two loosely coupled probes. In the terminology of four–terminal measurements the resistance given by (5) is equal to $\mathcal{R}_{12,34}$, where the first two indices indicate carrier source and sink and the second pair indicates the probes used to measure voltages. In the absence of a magnetic field reciprocity requires [7,11,12]

$$\mathcal{R}_{12,34} = \mathcal{R}_{34,12} . \qquad (6)$$

We could equally well inject and remove current through the weakly coupled probes and use the strongly coupled probes 1 and 2 to measure voltages. Reciprocity implies imediately, that we should not make special assumptions for certain probes only because they are in one particular configuration used as voltage probes.

For completeness, we also consider the "three terminal" contact resistances, $\mathcal{R}_{12,13} \equiv \mathcal{R}_{\text{con,L}} = (\mu_1-\mu_3)/eI$ and $\mathcal{R}_{12,42} \equiv \mathcal{R}_{\text{con,R}} = (\mu_4-\mu_2)eI$. Using (2) and (3) and $I = (e/h)T(\mu_1-\mu_2)$ yields,

$$\mathcal{R}_{\text{con,L}} = \frac{h}{e^2} \frac{1}{T} \frac{T_{32}}{T_{31}+T_{32}} , \qquad (7)$$

$$\mathcal{R}_{\text{con,R}} = \frac{h}{e^2} \frac{1}{T} \frac{T_{41}}{T_{41}+T_{42}} , \qquad (8)$$

Additivity of voltages implies that the two terminal resistance [7,11,13,14] $\mathcal{R}_{12,12} = (h/e^2)1/T$ is equal to the sum of the two three terminal resistances given above and the four–terminal resistance (5), i.e. in the notation introduced above, $\mathcal{R}_{12,12} = \mathcal{R}_{\text{con,L}} + \mathcal{R} + \mathcal{R}_{\text{con,R}}$.

Eq. (1) has been applied rather successfully to a number of transport problems. The sensitivity of voltage measurements to the electron phase was pointed out [7] and observed [8] for the Aharonov–Bohm effect in small loops in a four–terminal configuration. Eq. (1) provided a useful starting point for the development of theories to describe voltage fluctuations in small conductors [15]. Eq. (1) is also used to discuss the low magnetic field anomalies in ballistic conductors [16], and has been used to investigate the role of quantum coherence in the series addition of resistances applied to resonant tunneling [17] and to point contacts [18]. The success of (1) is particularly lucid in the description of the coherent injection and detection of carriers in ballistic electron focusing experiments [19]. It is theories which emphasize the balance of currents at measurement terminals which have led to a rather direct explanation of the quantum Hall effect and the prediction and observation of simultaneously quantized longitudinal and Hall resistances [20]. A formal derivation of (1) based on linear response formalism has also been achieved [21]. A continuum version of (1) describing a conductor with many closely spaced terminals appears in a theory which incorporates inelastic scattering into the conduction process [22].

We emphasize that the resistances as introduced above are not a consequence of the piling up of charge and depletion of charge near scattering objects. Such an approach was taken in [23] and has subsequently been further developed [24]. What matters, according to (1) are currents only. It is the probability of carriers to reach a certain terminal which determines the voltage. The formulae presented above are valid independent of the cross section of the conductor. Below we consider, for simplicity, an effective transmission problem with only two states at the Fermi energy.

The main effect which we analyze below results from the fact that a wave reflected from an obstacle has a definite phase–relationship to the wave incident on the obstacle. There is, therefore, in the arrangement of Fig.1 the possibility of detecting the interference of the incident wave with the reflected wave. More precisely, both the wave incident, say from the current source terminal 1 and the wave reflected from the barrier, contribute to the net probability for transmission into terminal 3. Thus T_{31} has, in general, terms which stem from the interference of the incident and the reflected wave. This term depends not only on the incident and reflected wave but also on the precise way these waves are coupled at the junction to terminal 3. These oscillations are a function of the distance of the probe away from the scattering obstacle. In the absence of phase randomizing events this dependence is oscillatory. The incident wave and the reflected wave are phase–coherent and contribute both to the net transmission probability from the conductor into the measurement probe. Based on (2) we can define three types of voltage measurements: A completely quantum mechanical calculation of the transmission probabilities including the interference terms described above is termed a *phase–sensitive* voltage measurement. A *phase–averaged* voltage measurement can be studied by assuming that we move the probe along the conductor and average over the resulting oscillatory chemical potential. A third type of voltage measurement is obtained by assuming that there is a physical, phase–randomizing agent which destroys the possibility of interference between the incident and reflected wave. We term such a voltage measurement *phase–insensitive*. In the latter case it is only currents, and not their quantum mechanical amplitude, which determine the measured voltage. Due to quantum mechanical interference effects these three voltage measurements will in general give rise to differing resistances [11]. The investigation of the resulting resistances is the main subject of this article.

2. CHEMICAL POTENTIAL WAVES

A. One–Parameter Coupler

Consider the simple example of a one dimensional conductor shown in Fig.1b. The arrows indicate the direction of motion of the carriers. Quantum mechanically, we describe carrier flow by current amplitudes. The square represents a scatterer which is specified by a 2×2–matrix with elements r and t for carriers incident from the left and r' and t' for waves incident from the right. We have $\alpha'_1 = r\alpha_1 + t'\alpha_2$ and $\alpha'_2 = t\alpha_1 + r'\alpha_2$. The triangle in Fig. 1b represents a splitter described by a 3×3–scattering matrix, which describes the connection of the incident waves to the outgoing waves [9,10]. We have $\beta'_i = \sum_j s_{ij}\beta_{ij}$ with j = 1,2,3. . Consider now a splitter which couples the measurement lead 3 only weakly to the conductor. Specifically, let $s_{31} = \epsilon^{1/2}$ and $s_{32} = \epsilon^{1/2}$. A fully specified splitter of this type, with only real matrix elements, is given in Ref. 25 and discussed in an Appendix. Later we consider more general couplers. Now let us calculate the transmission probability for carriers incident in lead 1 to arrive in lead 3. We have $\beta_1 = 1$ and the incoming amplitudes β_3 and α_2 are equal to zero. The amplitude β'_3 has two contributions: One contribution arises via direct coupling of the incident wave to lead 3 and the other contribution arises from carriers which pass the junction, and after reflection at the scatterer, are transmitted into lead 3. (Multiple scattering between the coupler and the scatterer can be neglected because of our assumption of weak coupling of the splitter to the conductor). Thus $\beta'_3 = s_{31} + s_{32}r = \epsilon^{1/2}(1+r)$ which gives a transmission probability $T_{31} = \epsilon|1+r|^2$. The transmission probability for carriers incident in lead two is obtained by taking $\alpha_2 = 1$ and $\beta_1 = 0$, $\beta_3 = 0$. In this case the wave emanating from the scatterer is t' and the amplitude in lead three is $\beta_3 = \epsilon t'$. Thus $T_{32} = \epsilon|t'|^2 = \epsilon T$. Here we have used the fact that for a 2×2 scattering matrix $t = t'$ and have expressed the square of the transmission amplitude in terms of the transmission probability T. For the reflection coefficient, we put $r = R^{1/2} \exp(i(\Delta\phi+2\phi))$, where $\Delta\phi$ is the phase change associated with reflection at the scatterer, and $\phi = k_F d_L$ is the phase accumulated traversing the distance d_L between the scatterer and the splitter. Hence $T_{31}= \epsilon(1+R+2R^{1/2}\cos\chi_3)$ where $\chi_3 = (\Delta\phi+2\phi)$ denotes the total relevant phase. Using (2) we find that the measured electro–chemical potential at probe 3 is,

$$\mu_3 = \mu_2 + \tfrac{1}{2}\frac{(1+R+2R^{1/2}\cos\chi_3)}{1+R^{1/2}\cos\chi_3}(\mu_1-\mu_2). \tag{9}$$

First we note that μ_3 is independent of the coupling parameter ϵ. There is no information left about the matrix describing the coupling of the conductor to the measurement lead. This is due to our particular specification of the scattering matrix. Later on we consider more general couplers. The chemical potential is nevertheless sensitive to a phase. This is a consequence of interference of the incident wave with itself. The phase which is accumulated traversing the distance d_L between the scatterer and the probe is $\phi = k_F d_L$, where k_F is the Fermi wave vector.

Let us now imagine that we can vary the position of the probe, as would be possible with a tunneling microscope [26,27]. There are tunneling tips where the current from the tip to the sample essentially flows through a single protruding adatom [27]. Alternatively, conductors which are gated offer the possibility of keeping the geometry fixed and changing the phase ϕ by changing the gate voltage. Instead of a tunnel microscope, split gates could be used to obtain loosely coupled probes. The work of van Houten et al. [19] shows that it is possible to operate such gates in a regime where their width is less than a Fermi wave–length. Due to the dependence on ϕ the measured chemical potential oscillates with a period $\lambda_F/2$. In the case of metallic conductors this period is of atomic dimensions but in ballistic conductors this period can be several hundred Angström. Oscillations with period $\lambda_F/2$ are familiar from Friedel's analysis of the equilibrium electro–static screening potential of impurities [28]. The oscillations in the chemical potential can thus be viewed as a non–equilibrium analogy of the Friedel oscillations. However this analogy is rather limited: in a many channel conductor, as discussed briefly at the end of this paper, the oscillations in the chemical potential can be of a much longer period and are not sinusoidal.

The chemical potential (9) is minimal for $\chi_3 = -1$ and maximal for $\chi_3 = 1$. From (9) we obtain,

$$\mu_{3\pm} = \mu_2 + \tfrac{1}{2}(1\pm R^{1/2})(\mu_1-\mu_2), \tag{10}$$

where the plus sign yields the maximum and the minus sign yields the minimum. From (10) we see that the oscillation amplitude is proportional to the square root of the reflection coefficient and is proportional to the chemical potential difference between the current source and sink,

$$\Delta\mu = R^{1/2}(\mu_1 - \mu_2). \tag{11}$$

In a more realistic discussion phase randomization and or inelastic scattering needs to be taken into account. In this case we can expect that the oscillations decay as the measurement lead is moved away from the scatterer. Eventually, if the measurement lead is more than an inelastic or phase breaking length away from the scatterer the phase sensitivity is completely lost.

For the probe to the right of the scatterer we find

$$\mu_4 = \mu_2 + \tfrac{1}{2}\frac{T}{1+R^{1/2}\cos(\chi_4)}(\mu_1 - \mu_2). \tag{12}$$

The pase $\chi_4 = (\Delta\phi' + 2\phi')$ contains a term proportional to the phase change for reflection from the right and a term for the phase $\phi' = k_F d_R$, which is a measure of the distance d_R of the probe from the scatterer. At first glance (12) looks rather different from (9), but surprisingly the maximum chemical potential and the minimum chemical potential are again given by (10), $\mu_{4pm} = \mu_{3pm}$. Surprisingly, the scatterer does not affect the magnitude of the chemical potential at all! But the scatterer has an effect on the phase and on the anharmonic content of the oscillations of the chemical potential. Thus, while the oscillations to the right and left of the scatterer are of equal magnitude, they are, in general, out of phase. The chemical potential, using (9) and (12) is shown in Fig.2. Fig. 2a depicts the case of a weak scatterer, $T = 0.9$, and an overall phase–change of π across the scatterer. The oscillations are sinusoidal with a very small anharmonic component superimposed. Fig. 2b shows a more opaque scatterer, $T = 0.5$, which exhibits increased amplitudes and increased anharmonic oscillations. Fig. 2c depicts the case of a strong scatterer, $T = 0.1$. The oscillations are strongly anharmonic; μ_4 exhibits Lorentzian spikes, whenever $\cos\chi_4 = -1$ with a height $(1/2)(1+R^{1/2})$ and a width $2^{3/2}(1-R^{1/2})$; μ_3 exhibits Lorentzian troughs with a depth and width equal to the height and width of the spikes of μ_4. From Fig.2 it is evident that the extremal values of the potential difference $\mu_3 - \mu_4$ are given by the amplitude of the chemical potential wave (11). The maximum positive potential drop is $\mu_3 - \mu_4 = R^{1/2}(\mu_1 - \mu_2)$ and the maximal negative potential difference is $\mu_3 - \mu_4 = -R^{1/2}(\mu_1 - \mu_2)$.

Let us next calculate the resistance of the scatterer. Using (9) and (12) we find

$$\mathcal{R} = \frac{h}{e^2}\frac{1}{2}\frac{2R+(1+R)R^{1/2}(\cos\chi_3+\cos\chi_4)+2R\cos\chi_3\cos\chi_4}{(1+R^{1/2}\cos\chi_3)(1+R^{1/2}\cos\chi_4)}. \tag{13}$$

For $\cos\chi_3 = \cos\chi_4 = 1$ this resistance is maximal and given by $\mathcal{R} = (h/e^2)R^{1/2}/T$ and for $\cos\chi_3 = \cos\chi_4 = -1$ this resistance is minimal and given by $\mathcal{R} = -(h/e^2)R^{1/2}/T$. Due to the oscillatory nature of the chemical potential the four–terminal resistance can be negative [11,19]. The negative four–terminal resistance does not violate any fundamental principle. The overall dissipation (to lowest order in ϵ) is given by the two–terminal resistance and equal to $W = (1/h)T(\mu_1 - \mu_2)^2$. Fig.3 shows the resistance (13), for $T = 0.5$ as a function of χ_4 for fixed values of $\cos\chi_3 = 1$ (top curve), $\cos\chi_3 = 0$ (middle curve), and $\cos\chi_3 = -1$ (bottom curve).

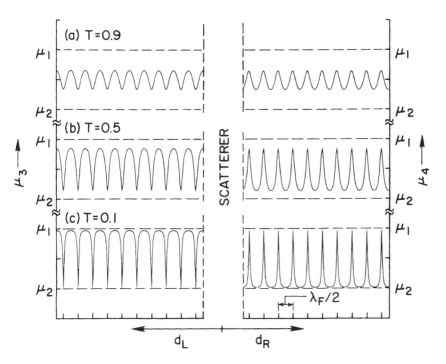

Fig. 2   Chemical potential oscillations as a function of the distance d_L and d_R of the probes away from the scatterer for three transmission probabilities, $T = 0.9$, $T = 0.5$, and $T = 0.1$.

Using (9) and (12) we can calculate the three–terminal contact resistances (7) and (8). We find,

$$\mathcal{R}_{con,L} = \frac{h}{2e^2} \frac{1}{1+R^{1/2}\cos\chi_3}, \tag{14}$$

and

$$\mathcal{R}_{con,R} = \frac{h}{2e^2} \frac{1}{1+R^{1/2}\cos\chi_4}, \tag{15}$$

Depending on the phase χ_3 or χ_4, these phase–coherent contact resistances can be as small as $(h/e^2)/(1 + R^{1/2})$ or as large as $(h/e^2)/(1 - R^{1/2})$. In particular, it is entirely possible that the three terminal resistances exceed h/e^2 by orders of magnitude.

Suppose we move the probe along the conductor and calculate the average of the measured potential. The average chemical potential to the left is,

$$\langle\mu_3\rangle = \mu_2 + \tfrac{1}{2}(2-T^{1/2})(\mu_1-\mu_2) . \tag{16}$$

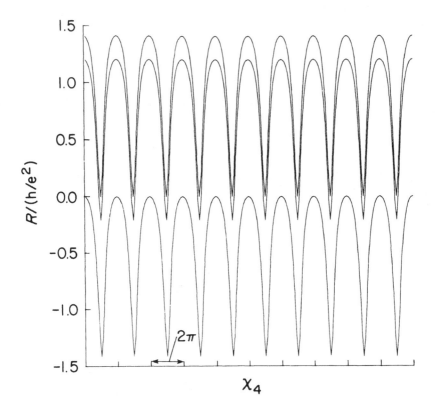

Fig.3   Phase–sensitive four–terminal resistance (13), as a function of the phase
         angle χ_4 for fixed values of $\cos\chi_3 = 1$, (top), 0 (middle) and -1 (bottom
         curve).

Similarly, the average potential to the right is

$$\langle\mu_4\rangle = \mu_2 + \tfrac{1}{2}T^{1/2}(\mu_1-\mu_2) \ . \tag{17}$$

Thus the average phase–sensitive resistance is

$$\langle\mathscr{R}\rangle = \frac{h}{e^2}\frac{1-T^{1/2}}{T} \ . \tag{18}$$

We see that it is upon phase–averaging that we obtain a positive resistance. The
phase averaged contact resistances are

$$\langle\mathscr{R}_{\mathrm{con,L}}\rangle = \langle\mathscr{R}_{\mathrm{con,R}}\rangle = \frac{h}{e^2}\frac{1}{2}\frac{1}{T^{1/2}} \ . \tag{19}$$

We can, as is often done in the literature [15], consider the mean square deviations
away from the ensemble averaged potentials and resistances. (This is an

appropriate characterization of the fluctuations only if the distribution functions are well approximated by a Gaussian function). With the abbreviation $eV = (\mu_3 - \mu_4)$ and (9), (12), (16), and (17) we find, after performing the integrals,

$$<(eV-<eV>)^2> = (1/2)\,T^{1/2}(1-T^{1/2})(\mu_1-\mu_2)^2. \tag{20}$$

The resistance fluctuations are defined by $\Delta\,\mathcal{R} = \mathcal{R} - <\mathcal{R}> = (V-<V>)/I$. With the help of (20), we find

$$<(\Delta\,\mathcal{R})^2> = (\frac{h}{e^2})^2\,\frac{1}{2}\,\frac{(1-T^{1/2})}{T^{3/2}}. \tag{21}$$

The mean square resistance fluctuations tend to zero as T approaches one and diverge as T approaches zero. This shows that the fluctuations in the measured resistance due to quantum interference effects at the measurement probe are not small even for a most ideal probe. Similarly, the conductance fluctuations can be expected to be very large in the limit of a very transparent scatterer. Here we have averaged the chemical potentials, or equivalently, the resistance. We could have focused on the conductance instead. We only stress that the averaged conductance is not the inverse of the averaged resistance.

Let us now discuss an additional measurement. Suppose that the phase of the wavefunction is disrupted before a carrier reflected from the barrier reaches the probe. More precisely, we assume that there are events which destroy the phase–coherence but leave the momentum of the carriers unchanged. Such a disruption of the phase requires a physical agent, such as magnetic impurities or low energy (acoustic) phonons, which cause small energy transfers but leave the momentum of the carriers unaffected. In such a situation, we do not have to treat the currents at the junction quantum mechanically. We can match currents instead of current amplitudes. This is equivalent to neglecting the phase dependent terms in (9) and (12). Alternatively, we can average the numerator and the denominator separately. This yields $\mu_3 = \mu_2 + (1/2)(1+R)(\mu_1-\mu_2)$ and $\mu_4 = \mu_2 + (1/2)(1-R)(\mu_1-\mu_2)$. We find a resistance

$$\mathcal{R} = \frac{h}{e^2}\,\frac{1-T}{T}, \tag{22}$$

i.e. a resistance proportional to $(1-T)/T$ as found by Landauer [5]. This resistance is larger than the averaged phase sensitive resistance. The results are identical in the limit of a completely transparent scatterer and in the limit of a completely opaque scatterer. We emphasize that the average taken in (18) corresponds to a special ensemble. It is obtained by varying the position of the voltage probes, but the conducting channel remains unaffected. The results of this section are summarized in Fig. 4. This figure shows the potential drop across the scatterer divided by the potential difference between current source and current sink as a function of the transmission probability of the scatterer for the voltage measurements discussed above. The phase–insensitive voltage measurement, leading to (22), gives rise to the dashed and dotted line. The phase–averaged measurement, leading to (18), is associated with a smaller potential drop, and is indicated in Fig.4 by a solid line. The dotted line is the root mean square chemical potential fluctuations (20), away from the phase averaged value. The dashed line gives the maximum amplitude

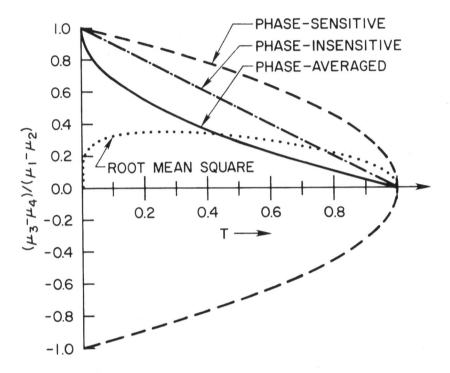

Fig.4   Potential drop as a function of transmission probability for the
             phase–averaged measurement (solid line), the phase–insensitive
             measurement (dashed and dotted line). A phase–sensitive measurement
             is bounded by the dashed line. Also shown are the root mean square
             voltage fluctuations for the phase–sensitive voltage measurement.

for the chemical potential difference for a phase–coherent measurement. A
phase–coherent voltage measurement gives a value for the potential drop which lies
anywhere in the region enclosed by the broken and dotted line, $\mu_3 - \mu_4 = \pm R^{1/2}(\mu_1-\mu_2)$.

All the results discussed above have been gained by considering a special
coupler describing the junction between probes 1 and 3 and 4 and 2. Below we
discuss some alternative couplers and discuss how these affect the results presented
so far. The terminal junction discussed above obeys $|s_{31}|^2 = |s_{32}|^2$, i.e. the splitter
couples with equal probability to carriers approaching the junction from the left
and from the right. This is not a fundamental symmetry of nature and can,
therefore, be broken.

B. Broken Left–Right–Symmetry

Consider a junction which does not couple equally to carriers arriving from the left
and right. We express this asymmetry by introducing an angle θ and put $s_{31} = (2\epsilon)^{1/2} \sin\theta$ and $s_{32} = (2\epsilon)^{1/2} \cos\theta$. For $\theta = \pi/4$ coupling to left and right moving
carriers is equal. A complete coupling matrix, valid for all θ, is specified in the

Appendix. For $\theta = 0$ such a measurement lead only couples to carriers approaching the junction from the right. For $\theta = \pi/2$, the junction only couples to carriers incident from the left. The transmission from branch 1 to branch 2 across such a junction is, however, close to unity only if the asymmetry parameter is close to $\pi/4$. It is this limit which we discuss. With these coupling elements we find to lowest order in ϵ, $T_{31} = |(2\epsilon)^{1/2}\sin\theta + (2\epsilon)^{1/2}r\cos\theta|^2$ and $T_{32} = |(2\epsilon)^{1/2}t\cos\theta|^2$ and hence, using (1),

$$\mu_3 = \mu_2 + \frac{\sin^2\theta + R\cos^2\theta + 2R^{1/2}\sin\theta\cos\theta\cos\chi_3}{1 + 2R^{1/2}\sin\theta\cos\theta\cos\chi_3}(\mu_1 - \mu_2). \qquad (23)$$

The ensemble averaged voltage (with respect to χ) is

$$\mu_3 = \mu_2 + \frac{(1 - R\sin^2 2\theta)^{1/2} - T\cos^2\theta}{(1 - R\sin^2 2\theta)^{1/2}}(\mu_1 - \mu_2). \qquad (24)$$

If the lead characterized by the same asymmetry parameter is moved to the right of the barrier we find $T_{41} = |(2\epsilon)^{1/2}t\sin\theta|^2$ and $T_{42} = |(2\epsilon)^{1/2}r'\sin\theta + (2\epsilon)^{1/2}\cos\theta|^2$ and hence,

$$\mu_4 = \mu_2 + \frac{T\sin^2\theta}{1 + 2R^{1/2}\sin\theta\cos\theta\cos\chi_4}(\mu_1 - \mu_2). \qquad (25)$$

The phase–averaged resistance is thus

$$<\mathcal{R}> = \frac{h}{e^2}\frac{1}{T}\frac{(1 - R\sin^2 2\theta)^{1/2} - T}{(1 - R\sin^2 2\theta)^{1/2}}. \qquad (26)$$

For the completely symmetric coupler, i.e. $\theta = \pi/4$, (18) is recovered. With increasing asymmetry the resistance, (26), tends towards the Landauer result (22). The effect of the asymmetry is to diminish the importance of the interference of the incident wave with the reflected wave.

The asymmetry of the coupler manifests itself in the contact resistances between the measurement probes and the reservoir. We find for the phase–averaged contact resistance to the left,

$$<\mathcal{R}_{con,L}> = \frac{h}{e^2}\frac{\cos^2\theta}{(1 - R\sin^2 2\theta)^{1/2}}, \qquad (27)$$

and to the right

$$<\mathcal{R}_{con,R}> = \frac{h}{e^2}\frac{\sin^2\theta}{(1 - R\sin^2 2\theta)^{1/2}}. \qquad (28)$$

For the phase–insensitive voltage measurement with identical probes to the left and right, we find again a resistance given by (22), independent of the asymmetry parameter θ. The phase–insensitive contact resistances, however, do depend on θ and are given by $\mathcal{R}_{con,L} = (h/e^2)\cos^2\theta$ and $\mathcal{R}_{con,R} = (h/e^2)\sin^2\theta$.

We have considered the case that the probes to the left and right are described by the same asymmetry parameter. For this case, in the absence of a barrier, $R = 0$, the four–terminal resistance is zero. If we were to choose differing asymmetry parameters to the left and right this would not be the case. Even a "perfect" conductor exhibits then a non–zero *four–terminal* resistance. Take $T = 1$, $(R = 0)$ in (23,25), and suppose the asymmetry is $\theta = \theta_L$ to the left and $\theta = \theta_R$ to the right. This gives a non–zero four–terminal resistance $(e^2/h)(\sin^2\theta_L - \sin^2\theta_R)$, which can be either positive or negative depending on the asymmetry parameters.

C. Sensitivity to Complex Matrix Elements

Our discussion so far has invoked a junction between the conductor and the probe which is described by *real* transmission amplitudes. This is an arbitrary restriction which is unlikely to describe an experimental junction. A generalization of the junction used above to a junction described by complex transmission amplitudes can be obtained by multiplying the real amplitudes s_{ij} by $\exp(i(\delta_1+\delta_j))$. Here, the δ_i are (arbitrary) phases. We obtain a scattering matrix with complex elements which satisfies the constraints imposed by current conservation and time reversal invariance. To investigate the effect of such phase factors let us calculate T_{31}. The incident wave is given by $\beta_1 = 1$. Direct reflection into probe 3 is proportional to s_{31} and hence we obtain an amplitude $\exp(i(\delta_1+\delta_2))(2\epsilon)^{1/2}\sin\theta$ contributing to β'_3. A second contribution to β'_3 comes from carriers which pass first through the junction, are reflected at the barrier and upon returning transmit into probe 3. The carriers passing through the junction pick up a factor $\exp i(\delta_1+\delta_2)$ at the junction and, upon returning from the barrier, have acquired a total phase factor $\exp i(\delta_1+\delta_2+2\phi+\Delta\phi)$. Upon transmission into probe 3, an additional phase proportional to $\delta_2+\delta_3$ is added. Thus these carriers contribute to β'_3 with an amplitude $\exp(i(\delta_1+2\delta_2+\delta_3+2\phi+\Delta\phi))(2\epsilon)^{1/2}\cos\theta$. Thus the transmission probability into probe 3 is $T_{31} = 2\epsilon\,|\sin\theta+\exp(2i\delta_2)r\,\cos\theta|^2$. The relevant phase in (23) is now $\chi_3 = 2\delta_2+\Delta\phi+2\phi$. A phase–sensitive measurement depends not only on the absolute values of the matrix elements which describe the coupling of a probe to a conductor, but also on the phases of these coupling elements. If the average is taken as above by moving the probes along the conductor, we again find a phase–averaged resistance given by (26), and a phase–insensitive resistance given by (22).

3. DISCUSSION

The oscillatory nature of the measured chemical potential is a quantum interference effect. In the simple geometries discussed above it is a result of the interference of the reflected wave with itself at the point of measurement. This quantum interference effect causes drastic departures from a more traditional picture, in which voltage drops across potential barriers are extended at most a few screening lengths away from the scatterer. As a result of quantum interference, even the most simple geometries can lead to longitudinal resistances which are not positive. To obtain a resistance with a definite sign some kind of phase averaging is needed or a process which destroys the phase–sensitivity of the voltage measurement process. In experiments on small conductors negative four–terminal resistances are now routinely observed (See Ref. 11 and 19 as guide to the literature).

There are a number of additional consequences of the results presented here.

The chemical potential drop across a sequence of scatterers can, as a consequence
of quantum interference, show little resemblance to the strength and distribution
of the scatterers. If phase–sensitivity is retained, the voltage does not necessarily
drop most strongly across the most opaque barriers. Another interesting question
concerns screening. To what extent can the electro–static potential follow the
chemical potential discussed above? The electric potential, as obtained from a
solution of Poisson's equation depends on the piled up charges in contrast to the
chemical potential discussed here which depends on the current amplitudes [11].
Obviously, if the screening length exceeds the Fermi wave–length, the electric
potential will exhibit a smoother behavior than the chemical potential. In any
case, for the electric potential we also expect the phase–sensitive, local carrier
densities to be the relevant starting point [11].

The single channel case provides a most dramatic illustration of the effects
discussed above. In a many channel conductor, differing channels give oscillatory
contributions to the chemical potential which are not in phase and this can lead to
a potential with an oscillation amplitude, which compared to the potential
difference between the current source and sink is smaller than in the single channel
case. The many channel case is, however, interesting. The oscillations are not only
determined by the wavelength of the individual quantum channels, but also by
sums and differences of these. This gives rise to a chemical potential which exhibits
oscillations not only with the period of a Fermi wave–length, but fluctuates on a
much larger length scale. Furthermore, we can have a large variety of weak
coupling probes. If we assume probes which do not couple equally to all the
channels, but provide for a spatial resolution also along the transverse direction of
the conductor, we can expect a most irregular potential with fluctuations as
dramatic as in the single channel case. On the other hand, for special junctions
which couple the quantum channels in differing branches in an irregular way, the
interference between incident and reflected waves can be avoided all together (such
a junction was used in Ref. 17).

We have emphasized voltage measurements with probes which permit an
exchange of carriers through the probe. Voltage measurements which avoid the
exchange of carriers are also possible [29]. To be of interest such measurement
procedures must couple locally to a conductor and if possible on a length scale
small compared to the phase–coherence length. It follows that the interference
pattern and the distinction which we illustrated here between phase–sensitive,
phase–averaged and phase–insensitive voltage measurements is likely to be of
importance regardless of the specific technique invoked.

Note added in proof. A brief account of the work reported here is given in Ref.30.

APPENDIX

For completeness we specify here the scattering matrices used in the text to
describe the junctions of the conductor of Fig.1 At the junction three incoming
current amplitudes β_i are linearly related to three out–going current amplitudes β'_i
by a matrix s_{ij}. Current conservation and time reversal require in the absence of a
magnetic field that s is unitary and symmetric. According to [9] the s matrix obeys
the following additional conditions: *Weak coupling* requires small transmission into
probe 3, $|s_{31}| = |s_{32}| = (\epsilon)^{1/2}$, where $\epsilon \langle\langle 1$ is the coupling parameter. The probe
has to be *non–invasive*, i.e. leave the transmission from terminal 1 to terminal 2
essentially unaffected, and Ref. 9 requires $|s_{21}| = |s_{12}| = 1 + O(\epsilon), |s_{11}| = |s_{22}| = O(\epsilon)$. Weak coupling by itself does not ensure that the probe is non–invasive.

Moreover, since we are concerned with phase–sensitive measurements, the conditions of Ref. 9 which are sufficient to guarantee that a phase–insensitive measurement is non–invasive do not guarantee that a phase–sensitive measurement is also non–invasive. This is because the conditions given above only determine the amplitudes but do not determine the phase of the matrix elements of s. With this in mind let us now present our examples of junction matrices.

1. The first example taken from Ref. 25 is a matrix with real elements s which is also symmetric with regard to the branches 1 and 2,

$$
\begin{bmatrix}
c & d & (\epsilon)^{1/2} \\
d & b & (\epsilon)^{1/2} \\
(\epsilon)^{1/2} & (\epsilon)^{1/2} & a
\end{bmatrix}
$$

This matrix is also unitary if, $a = \pm (1-2\epsilon)^{1/2}$, $d = - \epsilon/(b+c)$, and $b = c = \pm 1/2(-a\pm 1)$. The weak coupling solution which is also non–invasive if, $a = -(1-\epsilon+O(\epsilon^2))$, $b = -(\epsilon/2+O(\epsilon^2))$ and $d = +(1-\epsilon/2+O(\epsilon^2))$. The solution $a = (1-\epsilon+O(\epsilon^2))$, $b = (1-\epsilon/2+O(\epsilon^2))$ and $d = -(1-\epsilon/2+O(\epsilon^2))$ also describes a weak coupler but one which is invasive: Carriers transmitting the junction from branch 1 to 2 acquire a phase–shift of π. This changes the measured chemical potential as discussed in Sec.2C.

2. A coupling matrix which breaks the symmetry between the branches 1 and 2 with real matrix elements is given by

$$
\begin{bmatrix}
c & d & (2\epsilon)^{1/2}\sin\theta \\
d & b & (2\epsilon)^{1/2}\cos\theta \\
(2\epsilon)^{1/2}\sin\theta & (2\epsilon)^{1/2}\cos\theta & a
\end{bmatrix}
$$

where θ is the asymmetry parameter. This matrix is unitary if, $a = \pm(1-2\epsilon)^{1/2}$ as above, and $d = - 2\epsilon\sin\theta\cos\theta/(b+c)$, $c+b = \pm(-a\pm 1)$, and $c - b = 2\epsilon(\cos^2\theta - \sin^2\theta)/(c+b)$. This matrix is symmetric for $\theta = \pi/4$. For large asymmetries, i.e. if θ deviates considerably from $\pi/4$ the reflection of carriers incident in branch 1 (or 2) back into branch 1 (or 2) becomes large. The transmission from branch 1 to branch 2 is almost unity only for $|\theta-\pi/4| \langle\langle 1$. Note that even in this case, there is residual scattering even if the probe is completely decoupled $\epsilon = 0$. This junction, therefore, does not obey the conditions set forth in Ref. 9 discussed above. Note, that such residual scattering is not unphysical: If weak transmission into the measurement probe is provided by a wire with a barrier introduced by a narrow gate or by a split gate, the probe provides a perturbation of the wire geometry even if transmission into the probe is completely shut off. The weak coupling solution which is also (almost) non–invasive is $a = - (1-\epsilon+O(\epsilon^2))$, $c+b = - (1-\epsilon+O(\epsilon^2))$, $c-b = - 2\cos2\theta(1-\epsilon/2+O(\epsilon^2))$, and $d = \sin2\theta(1-\epsilon/2+O(\epsilon^2))$ with $|\theta-\pi/4| \langle\langle 1$.

3. As mentioned in the text these real scattering matrices can be made complex by multiplying each row and column with a factor $\exp(i\delta_j)$, where δ_j, $j = 1,2,3$ are phases. The junction is non–invasive if the phase change for transmission from 1 to 2 (or 2 to 1) is small. This is the case if $\delta_1+\delta_2$ is close to a multiple of 2π. This condition leaves open the possibility that the individual phases δ_1 and δ_2 are large (comparable to π) and hence have a considerable effect on the total transmission probability (see Sec.2C).

REFERENCES

[1]   J.R. Oppenheimer, Phys. Rev. 31 (1928) 66.
[2]   R.H. Fowler and L. Nordheim, Proc. Roy. Soc. (London) A119 (1928) 173.
[3]   J. Frenkel, Phys. Rev. 36 (1930) 1604; see also W. Ehrenberg and H. Hönl,
      Z. f. Physik 68 (1931) 289.
[4]   R. Landauer, IBM J. Res. Develop. 1 (1957) 223.
[5]   R. Landauer, Phil. Mag. 21 (1970) 863; and unpublished (1960).
[6]   P. Erdös and R.C. Herndon, Adv. Phys. 31 (1982) 65; P. Erdös, IBM Zürich
      research, unpublished (1967).
[7]   M. Büttiker, Phys. Rev. Lett. 57 (1986) 1761.
[8]   A.D. Benoit, S. Washburn, C.P. Umbach, R.B. Laibowitz, and R.A. Webb,
      Phys. Rev. Lett. 57 (1986) 1765.
[9]   H.L. Engquist and P.W. Anderson, Phys. Rev. B24 (1981) 1151.
[10] J.C. Slater, Microwave Electronics (Van Nostrand Toronto, New York,
      1950); D.M. Kerns and R.W. Beatty, Basic Theory of Waveguide Junctions
      and Introductory Microwave Network Analysis (Pergamon Press, Oxford,
      London, 1967).
[11] M. Büttiker, IBM J. Res. Develop. 32 (1988) 317.
[12] G.F.C. Searle, The Electrician 66 (1911) 999.
[13] Y. Imry, in Directions in Condensed Matter Physics, eds. G. Grinstein and
      E. Mazenko (World Scientific Press, Singapore, 1986) 101; A.G. Aronov and
      Yu. V. Sharvin, Rev. Mod. Physics 59 (1987) 755.
[14] D.S. Fisher and P.A. Lee, Phys. Rev. B23 (1981) 6851.
[15] M. Büttiker, Phys. Rev. B35 (1987) 4123; H.U. Baranger, A.D. Stone and
      D.P. DiVincenzo, Phys. Rev. B37 (1988) 6521; D. DiVincenzo and C.L.
      Kane, Phys. Rev. B38 (1988) 3006.
[16] G. Kirczenow, Phys. Rev. Lett. 62 (1989) 1920; 62 (1989) 2993; D.G.
      Ravenhall, H.W. Wyld and R.L. Schult, Phys. Rev. Lett. 62 (1989) 1780; Y.
      Avishai and Y.B. Band, Phys. Rev. Lett. 62 (1989) 2527; H.U. Baranger and
      A.D. Stone, (unpublished).
[17] M. Büttiker, IBM J. Res. Develop. 32 (1988) 63; Phys. Rev. B33 (1986)
      3020.
[18] C.W.J. Beenakker, H. van Houten, Phys. Rev. B39 (1989) 10445.
[19] H. van Houten, C.W.J. Beenakker, J.G. Williamson, M.E.I. Broekaart, and
      P.H.M. van Loosdrecht, B.J. van Wees and J.E. Mooij, C.T. Foxon and J.J.
      Harris, Phys. Rev. B39 (1989) 8556.
[20] For a brief review of this interesting development see, M. Büttiker, in
      Nanostructure Physics and Fabrication, eds. M.A. Reed and W.P. Kirk
      (Academic Press, New York , 1989) 319.
[21] H.U. Baranger and A.D. Stone, (unpublished).
[22] S. Datta and M.J. McLennan, (unpublished).
[23] M.Büttiker, Y. Imry, R. Landauer, S. Pinhas, Phys. Rev. B31 (1985) 6207.
[24] U. Sivan and Y. Imry, Phys. Rev. B33 (1986) 551; O. Entin–Wohlman, C.
      Hartzstein, and Y. Imry, Phys. Rev. B34 (1986) 921.
[25] M. Büttiker, Y. Imry, M.Ya. Azbel, Phys. Rev. A30 (1984) 1982.
[26] P. Muralt, D.W. Pohl, and W. Denk, IBM J. Res. Develop. 30 (1986) 443;
      J.R. Kirtley, S. Washburn, and M.J. Brady, Phys. Rev. Lett. 60 (1988) 1546;
      I BM J. Res. Develop. 32 (1988) 414.
[27] N.D. Lang, Phys. Rev. B36 (1987) 8173; Phys. Rev. B37 (1988) 10395.

[28]   J. Friedel, Phil. Mag. Suppl. 3 (1954) 446; Nuovo Cimento Suppl. 7 (1958)
       287.
[29]   R. Landauer, IBM J. Res. Develop. 32 (1988) 306.
[30]   M. Büttiker, Phys. Rev. B40 (1989) 3409.

M. Büttiker is with IBM Thomas J. Watson Research Center, P.O. Box 217,
Yorktown Heights, New York 10598, USA.

QUANTUM POINT CONTACTS AND COHERENT ELECTRON FOCUSING

H. van Houten and C.W.J. Beenakker

The theory of quantum ballistic transport, applied to quantum point contacts and coherent electron focusing in a two−dimensional electron gas, is reviewed in relation to experimental observations, stressing its character of electron optics in the solid state. It is proposed that an optical analogue of the conductance quantization of quantum point contacts can be constructed, and a theoretical analysis is presented. Coherent electron focusing is discussed as an experimental realization of mode−interference in ballistic transport.

1. INTRODUCTION

Electron optics as an experimental discipline [1] can be traced back to Busch [2] who demonstrated the focusing action of an axial magnetic field on electron beams in vacuum. The similarity of the properties of ballistic conduction electrons in a degenerate electron gas and those of free electrons in vacuum suggests the possibility of electron optics in the solid state. Classical ballistic transport in metals, which has the character of geometrical optics, has been realized with the pioneering work of Sharvin [3] and Tsoi [4] on point contacts and electron focusing. This work has recently been extended to the quantum ballistic transport regime in the two–dimensional electron gas[1] (2DEG) in GaAs–AlGaAs heterostructures [6–8]. Essential advantages of this system are its reduced dimensionality and lower electron gas density, with correspondingly large Fermi wavelength, which can be varied locally by means of gate electrodes. Point contacts of variable width of the order of the Fermi wavelength have been defined by applying a negative voltage to a split–gate on top of the heterostructure. An interesting and unexpected finding was the quantization of the conductance of these *quantum point contacts* in units of $2e^2/h$ [6,7] (see Fig.1). This new effect can be understood by the similarity of transport through the point contact and propagation through an electron waveguide. Point contacts can also be used to inject a divergent beam of ballistic electrons in the 2DEG. The focusing of such a beam by a transverse magnetic field, acting as a *lens*, has been demonstrated in an electron focusing experiment [8,9].

---

[1]A 2DEG is a degenerate electron gas which is strongly confined in one direction by an electrostatic potential well at the interface between two semiconductors, such that only the lowest quantum level in the well is occupied. Electrons in a 2DEG are thus dynamically constrained to move in a plane, and accordingly there is a 2D density of states [5].

W. van Haeringen and D. Lenstra (eds.), Analogies in Optics and Micro Electronics, 203–225.
© 1990 Kluwer Academic Publishers. Printed in the Netherlands.

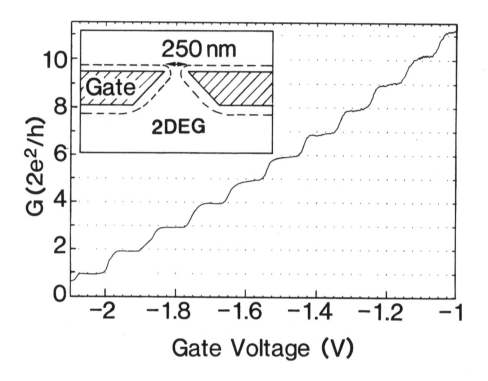

Fig.1   Conductance of a quantum point contact in the absence of a magnetic
          field. The inset shows the sample geometry schematically. The point
          contact is electrostatically defined as a constriction in the
          two–dimensional electron gas. The constriction width increases
          continuously as the (negative) gate voltage is decreased, but the
          conductance increases in steps given by $2e^2/h$. [From Ref. 6]

The boundary of the 2DEG, also defined by means of a gate, acts as a high quality
mirror, causing specular boundary scattering [8,9]. For sufficiently large gate
voltages the point contact width is smaller than the Fermi wavelength of the
electrons. Quantum point contacts in this regime act as monochromatic *point
sources* [10], as demonstrated by the large interference structure found
experimentally [8,9]. The first building blocks for the exploitation of solid state
electron optics have thus been realized.
     In this chapter we discuss the theory of quantum ballistic transport applied to
the conductance through quantum point contacts [6,7] and coherent electron
focusing [8–10], stressing its character of electron optics in the solid state. A review
with a wider scope and more experimental detail is Ref. 11. Reviews of related

topics in quantum ballistic transport can be found in Refs. 12–14. Section 4 of this paper, dealing with the optical analogue of the conductance quantization of a quantum point contact, does not contain previously published material.

2. ELECTRONS AT THE FERMI LEVEL

In this section we discuss some elementary properties of the single electron states in a degenerate two–dimensional electron gas. In a large system, and in the effective mass approximation, these states are plane waves with wave vector **k**. The conduction band is approximated by

$$E(\mathbf{k}) = \frac{\hbar^2 k^2}{2m} , \tag{1}$$

with m the effective mass, which for GaAs is 0.067 m_e. Linear transport at low temperatures can be formulated in terms of motion of electrons at the Fermi level E_F. The group velocity of these conduction electrons is $\mathbf{v}_F = \partial E / \partial \hbar \mathbf{k} = \hbar \mathbf{k}_F / m$, and their wavelength $\lambda_F = h / m v_F$ is ≈ 40 nm in the experiments. These quantities are obtained from the electron gas density $n_s = (k_F)^2 / 2\pi$, which is a directly measurable quantity.

In a channel with width comparable to λ_F the transverse motion of the electrons is quantized, while the motion along the channel is free. The wave function is separable in a transverse bound state with quantum number n, and a longitudinal plane wave $\exp(iky)$. The dispersion relation $E_n(k)$ of these quasi one–dimensional subbands or transverse waveguide modes is

$$E_n(k) = E_n(0) + \frac{\hbar^2 k^2}{2m} , \tag{2}$$

where $E_n(0)$ is the energy of the nth bound state. The frequency $E_n(0)/\hbar$ is the cut–off frequency of the mode, as in an optical fiber. The number of occupied modes N is the largest integer n such that $E_n(0) \le E_F$. Since E is still quadratic in k, the group velocity $\hbar k / m$ remains linear in the wave number, as in the bulk 2DEG.

This changes if we apply a magnetic field B perpendicular to the electron gas. Let the channel be in the y–direction, and B in the z–direction, so that the 2DEG is in the x–y plane. In the Landau gauge $\mathbf{A} = (0, Bx, 0)$ the wave function remains separable as in the absence of a magnetic field. However, $E_n(k)$ is no longer quadratic in k, so that the group velocity differs from $\hbar k / m$. Note that it is the group velocity which is relevant for conduction through the channel. In fact, the current carried by a mode n is proportional to the product of the group velocity $v_n = dE_n(k)/\hbar dk$ and the density of states ρ_n, both evaluated at the Fermi energy. Since the number of states per unit channel length in an interval dk is $2\, dk/2\pi$ (with an additional factor of 2 from the spin degeneracy) the energy density of states is $\rho_n = (\pi dE_n(k)/dk)^{-1}$. It follows that $\rho_n v_n = 2/h$ is *independent* of wave number or mode index, regardless of the form of the dispersion relation. Formulated differently, this tells us that in an electron waveguide the *current is shared equally among the modes*. This is the basic reason for the conductance quantization discussed in the following section.

3. CONDUCTANCE QUANTIZATION OF A QUANTUM POINT CONTACT

Van Wees *et al.* [6] and Wharam *et al.* [7] observed a sequence of steps in the
conductance G of a point contact as its width was varied (by means of a gate
voltage (see Fig.1). The steps were at integer multiples of $2e^2/h \approx (13\ k\Omega)^{-1}$, a
combination of fundamental constants which is familiar from the quantum Hall
effect [15]. However, the conductance quantization was observed in the *absence* of a
magnetic field, as well as in the presence of a magnetic field. An elementary
explanation of this effect relies on the fact that the point contact acts in a way as
an electron waveguide or multi–mode fiber [16,17]. Each populated
one–dimensional subband or transverse waveguide mode contributes $2e^2/h$ to the
conductance because of the cancellation of the group velocity and the
one–dimensional density of states discussed in sec. 2. Since the number N of
occupied modes is necessarily an integer, it follows from this simple argument that
the total conductance is quantized

$$G = \frac{2e^2}{h} N , \tag{3}$$

as observed experimentally[2].

For a square well confinement potential of width W one has $N = k_F W/\pi$ in zero
magnetic field, if $W >> \lambda_F$ so that the discreteness of N may be ignored. Eq. (3)
then gives the 2D analogue of the 3D Sharvin formula [3] for the conductance of a
classical ballistic point contact. However, the above argument leading to (3) holds
irrespective of the shape of the confinement potential, or of the presence of a
magnetic field.

A more detailed explanation of the conductance quantization requires a
consideration of the coupling between quantum states in the narrow point contact
to those in the wide 2DEG regions. As discussed in sec. 6 in the more general
context of electron focusing, this is essentially a transmission problem. Here it
suffices to say that if the point contact opens up gradually into the wide regions,
the transport is *adiabatic* [19–21] from the entrance to the exit of the point contact.
The lowest N subbands at the wide entrance region (in one–to–one correspondence
to those occupied in the point contact) are transmitted with probability 1, the
others being reflected.

For the case that the point contact widens *abruptly* into the 2DEG, deviations
from (3) occur, for short channels mainly because of evanescent waves (or modes
with a frequency above the cut–off frequency) which have a non–zero transmission
probability, for longer channels mainly because of quantum mechanical reflections
at their entrance and exit. Extensive numerical and analytical work [22–26] has
demonstrated that (3) is still a surprisingly good approximation –although
transmission resonances may obscure the plateaus. Experimentally, a limited
increase in temperature helps to smooth out the resonances and to improve the
flatness of the plateaus.

---

[2]The resistance $1/G$ is non–zero for an ideal ballistic conductor of finite width,
because it is a *contact* resistance. The existence of a contact resistance of order
h/Ne^2 was first pointed out by Imry [18].

4. OPTICAL ANALOGUE OF THE CONDUCTANCE QUANTIZATION

The analogy of ballistic transport through quantum point contacts with transmission of light through a multi mode fiber, or microwaves through a waveguide (summarized in Table I), naturally triggers the question: "Does the quantized conductance have an optical analogue, and if it does, why was it not known?" In this section we want to show that: "Yes there is an optical analogue, but the experiment is not as natural for light as it is for electrons."

Consider monochromatic light, of frequency ω and polarization $E_x = E_y = B_z = 0$, incident on a long slit (along the z-axis) in a metallic screen. The non–zero electric field component $E_z(x,y)$ then satisfies the two–dimensional scalar wave equation (Ref. 27, page 561)

$$\frac{\partial^2 E_z}{\partial x^2} + \frac{\partial^2 E_z}{\partial y^2} + k^2 E_z = 0, \tag{4}$$

with $k \equiv \omega/c \equiv 2\pi/\lambda$ the wave number (c is the velocity of light, and λ the wavelength). This equation is identical to the Schrödinger equation for the wave function $\Psi(x,y)$ of an electron at the Fermi level in a 2DEG, with the identification $k \equiv k_F$. If the boundaries in the 2DEG are modeled by infinite potential walls, then the boundary conditions are also the same in the two problems (Ψ and E_z both vanish for (x,y) on the boundary). From E_z one finds the non–zero magnetic field components $B_x = -(i/k)\partial E_z/\partial y$, and $B_y = -(i/k)\partial E_z/\partial x$, and hence the energy flux

$$\mathbf{j} = \frac{c}{8\pi} \operatorname{Re}(\mathbf{E} \times \mathbf{B}^*) = \frac{c}{8\pi k} \operatorname{Re}(i E_z \nabla E_z^*). \tag{5}$$

This expression is identical, up to a numerical factor, to the quantum mechanical expression for the particle flux, $(\hbar/m) \operatorname{Re}(i\Psi \nabla \Psi^*)$. It follows that the ratio of transmitted to incident power in the optical problem is the same as the ratio of the transmitted to incident current in its electronic counterpart.

In optics one usually studies the transmission of a single incident plane wave, as a function of the angle of incidence. In a 2DEG, however, electrons are incident from all directions in the x–y plane, with an isotropic velocity distribution. The incident flux then has a $\cos\theta$ angular distribution, where θ is the angle with the normal to the screen. [Such an isotropic distribution results naturally from a "good" electron reservoir, because it is in thermal equilibrium]. If an equivalent illumination can be realized optically, then the incident power is distributed equally among the 1–dimensional transverse modes in front of the screen –as in the electronic problem. The transmission probability T (which appears in the Landauer formula (11), see section 6) is defined as the ratio of the total transmitted power P_{trans} to the incident power per mode (both per unit of length in the z–direction). The latter quantity is $\lambda/2$ times the total incident energy flux j_{in}, so that

$$T \equiv \frac{2 P_{trans}}{j_{in}\lambda}. \tag{6}$$

Table I

photons	electrons
ray	trajectory
mode	subband
mode index	quantum number n
wave number k	canonical momentum $\hbar k$
frequency ω	energy $E = \hbar\omega$
dispersion law $\omega(k)$	bandstructure $E_n(k)$
group velocity $d\omega/dk$	group velocity $dE/\hbar dk$

The one–to–one correspondence with the electron transport problem now tells us that T is approximately quantized to the number N of 1D transverse modes in the slit,

$$T \approx N = \text{Int} \left[\frac{2W}{\lambda} \right] . \qquad (7)$$

On increasing the width W of the slit one would thus see a *step wise* increase of the transmitted power, if the incident flux is kept constant. The height of the steps is $j_m\lambda/2$. To observe well–developed plateaus the screen should have a certain thickness d, in order to suppress the evanescent waves through the slit. A thickness $d \approx (W\lambda)^{1/2}$ is sufficient (see Ref. 22). If d is much greater, the transmission steps can become obscured by Fabry–Perot like resonances from waves reflected at the front and back end of the slit. These can be avoided by smoothing the edges of the slit. Additional interference structure (not present in the electronic case) can occur if light incident from different angles is coherent.

Instead of the polarization given above, one can also use the polarization $B_x = B_y = E_z = 0$. The wave function Ψ then corresponds to the magnetic field component B_z. The boundary condition on B_z is that its normal derivative vanishes on the screen. Although it is not obvious how to realize this boundary condition in the electronic case, the transmission steps are probably present irrespective of the boundary conditions (the conductance quantization in a 2DEG occurs both for smooth and steep potential walls).

The above geometry was chosen to achieve a mapping onto the scalar two–dimensional electron transport problem. In 3D an electronic system showing the conductance quantization has not yet been realized experimentally, although it should be possible in principle. The 3D optical analogue may be more readily realizable. Consider a metallic screen in the x–y plane with an aperture of arbitrary shape. At a frequency ω a number N of 2–dimensional transverse modes in the aperture are below cut–off. We would expect a *step wise* increase in the transmitted power on increasing the size of the aperture (i.e. on increasing N), if the aperture is illuminated in such a way that the incident flux is distributed equally among the 2D transverse modes in front of the screen. The number dN of these modes per unit area with wave vectors within a solid angle $d\Omega$ is (including both polarizations)

$$dn = 2 \frac{dk_y}{2\pi} \frac{dk_z}{2\pi} = \frac{k^2}{2\pi^2} \cos\theta d\Omega, \tag{8}$$

where θ is the angle with the normal to the screen. It follows that (as in the 2D case discussed earlier) a $\cos\theta$ distribution of the incident flux has the required equipartition of power among the transverse modes. This angular distribution is realized e.g. by the light scattered diffusely from a surface in front of the screen. The height of the transmission steps should be the incident power per mode. Now, the total number n of 2D transverse modes per unit area is

$$n = \int\limits_0^{2\pi} d\phi \int\limits_0^{\pi/2} \frac{dn}{d\Omega} \sin\theta \, d\theta = \frac{2\pi}{\lambda^2}. \tag{9}$$

We therefore predict for the 3D case that the height of the steps in the transmitted power (for an energy flux j_{in} incident on the screen) will be $j_{in}\lambda^2/2\pi$, assuming that the two independent polarizations of the modes in the aperture can be resolved (the steps are a factor of two larger if this is not the case).

It would be interesting to carry out this analogy between optics and micro–electronics experimentally.

5. CLASSICAL ELECTRON FOCUSING

Ballistic transport is in essence a transmission problem. One example is the conductance of a quantum point contact, which is determined by the total transmission probability through the constriction, in analogy with the transmission through a waveguide (cf. secs. 3 and 4). More sophisticated transmission experiments can be realized if one point contact is used as a collector. Electron focusing in the transverse field geometry due to Tsoi [4], is one of the simplest realizations of such an experiment. (Longitudinal electron focusing [3] is not possible in 2D). In our experiments, two adjacent point contacts are positioned on the same 2DEG boundary. The boundary as well as both point contacts are defined electrostatically in the 2DEG by means of a gate electrode of suitable shape, see Fig.2. A constant current I_i is injected through one of the point contacts, and the voltage V_c on the second point contact (the collector) is measured. A transverse magnetic field is used to deflect the injected electrons, such that they propagate in *skipping orbits* along the 2DEG boundary from the injector towards the collector –provided the boundary scattering is specular. In Fig.2 the skipping orbits are illustrated. A magnetic field acts as a *lens*, focusing the injected electron beam, as is evidenced in Fig.2 by the presence of *caustics* or lines of focus (here the classical density of injected electrons is infinite). The focal points on the 2DEG boundary are separated by the classical cyclotron orbit diameter $2l_{cycl} = 2mv_F/eB$. The classical transmission probability from injector to collector has a peak if a focus coincides with the collector, which happens if the distance between the two point contacts $L = p \times 2l_{cycl}$, with $p = 1,2,\ldots$. As a result, the collector voltage as a function of magnetic field shows a series of peaks, called a focusing spectrum.

The classical transmission probability can be calculated straightforwardly [9], as in the 3D metal case [28]. A plot of the classical focusing spectrum assuming purely specular boundary scattering is given in Fig. 3a using the experimental

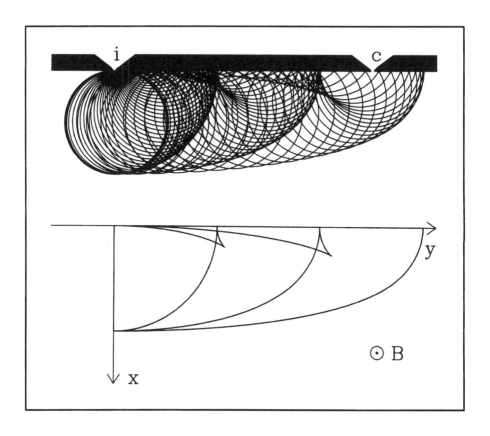

Fig.2   Skipping orbits at a 2DEG boundary. The gate defining the injector (i)
           and collector (c) point contacts and the boundary is shown
           schematically in black. For clarity the trajectories are drawn up to the
           third specular reflection only. Bottom: Calculated location of the caustic
           curves [From Ref.10]

parameters $L = 3.0$ μm and $k_F = 1.5 \times 10^8$ m^{-1} and an estimated injector and
collector width of 50 nm. The spectrum consists of a series of equidistant peaks of
constant amplitude at magnetic fields which are multiples of $B_{focus} \equiv 2\hbar k_F/eL \approx$
0.066 T. The focusing signal oscillates around a value given by the conventional
Hall resistance $B/n_s e$ seen in the reverse field signal. In reverse fields no focusing
peaks occur, because the electrons are deflected away from the collector. Note also
that V_c/I_i is approximately quadratic in B for weak positive fields in contrast to
the linear Hall resistance for reverse fields. These features are strikingly confirmed
by the experiment (see Fig. 3b), which constitutes a direct observation of skipping
orbits at the 2DEG boundary, and demonstrates that boundary scattering is highly
specular (because the peaks at higher magnetic fields, corresponding to electrons

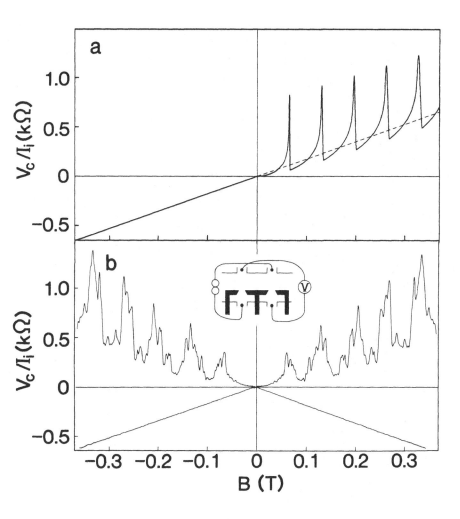

Fig.3   (a) Classical focusing spectrum, calculated with $W_i = W_c = 50$ nm, and $L = 3.0$ μm. The dashed line is the extrapolation of the classical Hall resistance seen in reverse fields. (b) Experimental electron focusing spectra obtained in the measurement configuration illustrated in the inset. The two traces have been obtained by interchanging current and voltage leads, and demonstrate the injector–collector reciprocity of (13). [From Ref.9]

multiply scattered by the boundary, do not diminish in amplitude). In contrast, in metals the amplitude of subsequent peaks usually diminishes rapidly, indicating partially diffuse boundary scattering [4,28]. Specular scattering occurs if the Fermi wavelength is large compared to the spatial scale of the boundary roughness. This is difficult to achieve in metals (where $\lambda_F \approx 0.5$ nm), but not in a 2DEG (where λ_F is 100 times larger). This explains why specular scattering is found to be predominant in our experiments [8,9].

While the overall shape of the focusing spectra is as one expects from the classical calculation, an additional unexpected oscillatory structure is observed in the experiment. This is the signature of a new phenomenon: coherent electron focusing [8]. As discussed in sec. 7, its origin is mode interference, the relevant modes being magnetic edge states coherently excited by the injecting point contact. Fig. 3b also shows the collector signal obtained after interchanging current and voltage leads, demonstrating the reproducibility of the fine structure. The symmetry of the focusing spectrum on interchanging injector and collector is an example of the Onsager–Casimir relation for the *conductance* (as opposed to conductivity) derived for the quantum ballistic transport regime by Büttiker [29] (see sec. 6). A related source–detector *reciprocity theorem* is well known in microwave transmission theory [30]. In optics this is known as the reciprocity theorem of Helmholtz [27].

6. ELECTRON FOCUSING AS A TRANSMISSION PROBLEM

Transport in the regime of diffusive motion is usually treated in terms of a local distribution function found from a self consistent solution of the linearized Boltzmann equation and the Poisson equation. Ballistic transport is inherently non–local, and can more naturally be viewed as a transmission problem. In the linear transport regime there is then no need to consider the self consistent electric field explicitly, which greatly simplifies the analysis. This approach originated in a paper by Landauer in 1957 [31], and has since been generalized and extended by Büttiker [29] to the case of a realistic conductor with multiple leads connected to the external current source and voltmeters. The equivalence of Büttikers approach and the more familiar linear response theory based on the Kubo formalism has been demonstrated [32–34]. In this section we apply the Landauer–Büttiker formula to the electron focusing geometry. For simplicity, we consider the three–terminal configuration of Fig.4, with point contacts in two of the probes, serving as injector and collector. (By a simple extension of the arguments in this section, one can also derive expressions for the four–terminal geometry of Fig. 3b [9]). In order to have well defined initial and final states for the scattering problem, the probes are connected via idealized leads (or electron waveguides) to reservoirs at a constant electro–chemical potential. We denote by μ_i and μ_c the chemical potentials of the injector and collector reservoirs. The chemical potential of the drain reservoir, which acts as a current sink, is assigned the value zero. Following Büttiker [29], we can relate the currents $I_\alpha (\alpha = i,c,d)$ in the leads to these chemical potentials via the transmission probabilities $T_{\alpha \rightarrow \beta}$ from reservoir α to β, and reflection probabilities R_α (from reservoir α back to the same reservoir). These equations have the form

$$\frac{h}{2e}I_\alpha = (N_\alpha - R_\alpha)\mu_\alpha - \sum_{\beta \neq \alpha} T_{\beta \rightarrow \alpha}\mu_\beta \ . \tag{10}$$

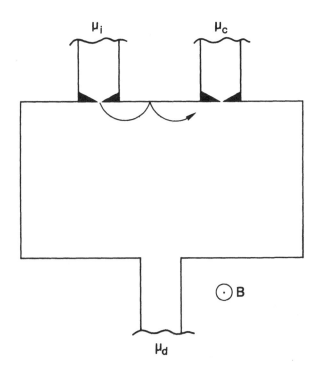

Fig.4   Three–terminal conductor in the electron focusing geometry. Three
reservoirs at chemical potentials μ_i, μ_c, and μ_d are connected by leads to
a wide 2DEG. Two of the leads contain a narrow constriction, or point
contact (shown in black). The current flows from reservoir i to reservoir
d, while reservoir c draws no net current. [From Ref. 9]

Here N_α is the number of occupied (spin degenerate) transverse waveguide modes
or "channels" in lead α. Eqs. (10) can also be used in the classical limit, where the
discreteness of N_α can be ignored. We first apply these equations to the
two–terminal conductance of a single point contact. In that case all $\mu_\beta = 0$ for
$\beta \neq \alpha$, and thus

$$G \equiv \frac{I_\alpha}{\mu_\alpha / e} = \frac{2e^2}{h} T, \tag{11}$$

with $T \equiv N_\alpha-R_\alpha$ the total transmission probability for the N_α channels in the leads. As discussed in sec. 3, for point contacts with $W \geq \lambda_F$ one has in a good approximation $T \approx N$, with $N \approx k_F W/\pi$ the number of occupied subbands in the constriction.

In the electron focusing experiments the collector is connected to a voltmeter, which implies $I_c = 0$ and $I_d = -I_i$. We then find from (10)

$$\mu_c = \frac{T_{i \rightarrow c}}{N_c-R_c} \mu_i,$$ (12a)

$$\frac{h}{2e}I_i = (N_i-R_i)\mu_i-T_{c \rightarrow i}\mu_c .$$ (12b)

The transmission probability $T_{c \rightarrow i} \approx 0$ because the magnetic field deflects electrons from the injector towards the drain, and away from the collector (see Fig. 4). The measured quantity is the ratio of collector voltage $V_c \equiv \mu_c/e$ (relative to the voltage of the drain) and injector current

$$\frac{V_c}{I_i} = \frac{2e^2}{h} \frac{T_{i \rightarrow c}}{G_i G_c},$$ (13)

where we have used (11). The transmission probability $T_{i \rightarrow c}$ is evaluated in the next sections. This result is symmetric under interchange of injector and collector leads, with simultaneous reversal of the magnetic field, because [29] $T_{i \rightarrow c}(B) = T_{c \rightarrow i}(-B)$. This symmetry relation explains the injector–collector reciprocity observed experimentally (see Fig. 3b).

7. COHERENT ELECTRON FOCUSING

7.1 Experiment

The oscillatory structure superimposed on the classical focusing peaks in the experimental trace of Fig. 3b is a quantum interference effect. At higher magnetic fields (beyond about 0.4 T), the collector voltage shows oscillations with a much larger amplitude than the low field focusing peaks, and the resemblance to the classical focusing spectrum is lost. The oscillatory structure becomes especially dramatic on decreasing the width of the point contacts (by increasing the gate voltage), and at very low temperatures, if also the voltage drop across the injector point contact is maintained below $k_B T/e$. In Fig. 5 we show the experimental results for a device with 1.5 μm point contact separation, and estimated point contact width of $20 - 40$ nm. Note the nearly 100% modulation of the focusing signal. A Fourier transform of the data (see inset of Fig.5) shows that the large–amplitude high–field oscillations have a dominant periodicity of 100 mT, approximately the same as the periodicity B_{focus} of the low field focusing peaks. This dominant periodicity is insensitive to changes in the gate voltage, and is the characteristic feature of coherent electron focusing which is most amenable to a direct comparison with theory.

In this section we present a theory of coherent electron focusing [9,10]. We start with a short discussion of magnetic edge states, which are the modes of this transmission problem. After a brief discussion of the excitation of these modes by

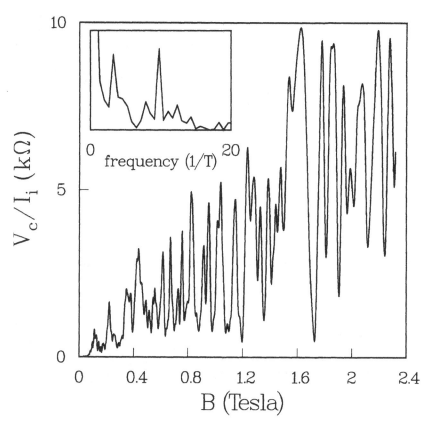

Fig.5   Electron focusing spectra measured at 50 mK for a device with 1.5 μm
          point contact separation, showing large quantum interference structure.
          The inset gives the Fourier transform power spectrum, for $B > 0.8$ T,
          demonstrating that the dominant periodicity is the low field classical
          focusing periodicity. [From Ref. 9]

the injecting point contact, we proceed by giving an explanation of the
characteristic features of the observed spectrum in terms of *mode interference*. This
treatment was first given in Ref. 10, and in a more detailed form in Ref. 9,
together with an equivalent one in the ray–picture (interference between
trajectories).

7.2 Skipping Orbits and Magnetic Edge States

The motion of conduction electrons at the boundaries of a 2DEG in a
perpendicular magnetic field is in skipping orbits, provided the boundary scattering
is specular (see Figs. 2, and 6). (In a narrow channel traversing states can coexist
with skipping orbits, as discussed *e.g.* in Ref. 16). The position (x,y) of the electron
on the circle with center coordinates (X,Y) can be expressed in terms of its velocity
by

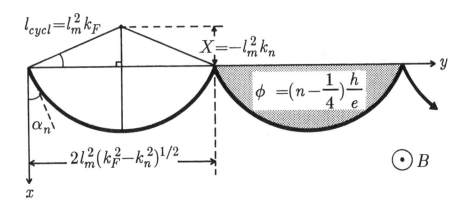

Fig.6   Skipping orbit at a 2DEG boundary. The semi–classical correspondence
         with a magnetic edge state is indicated. The flux ϕ enclosed in the
         shaded area equals $(n - \frac{1}{4})$ elementary flux quanta h/e.

$$x = X + v_y/\omega_c \ , \ y = Y - v_x/\omega_c \ , \tag{14}$$

with $\omega_c = eB/m$ the cyclotron frequency. The separation X of the center from the
boundary is constant on a skipping orbit, while Y jumps over a distance equal to
the chord length on each specular reflection. The canonical momentum $\mathbf{p} = m\mathbf{v} -
e\mathbf{A}$ is given in the Landau gauge $\mathbf{A} = (0,Bx,0)$ by

$$p_x = mv_x \ , \ p_y = -eBX \ , \tag{15}$$

so that p_y is a constant of the motion. The periodic motion perpendicular to the
boundary leads to the formation of discrete quantized states (with quantum
number n). These states are known as magnetic edge states (and in metals as
magnetic surface states [35,36]). The wave number *along* the boundary is a plane
wave with wave number $k_y \equiv p_y/\hbar$, which can be expressed in the guiding center
coordinate X according to $X = -(l_m)^2 k_y$. Here $l_m = (\hbar/eB)^{1/2}$ is the magnetic
length, which plays a role similar to that of λ_F in the absence of a magnetic field.
Coherent electron focusing constitutes an interference experiment with quantum
edge states at the Fermi energy. The wave number k_y for these states has a
quantized value, denoted by k_n. The phase of each edge state arriving at the
collector is determined by k_n, and the functional dependence of k_n on n is thus
important.

An exact quantum mechanical treatment is possible, but here it suffices to
consider the semi–classical Bohr–Sommerfeld approximation applied to the
periodic motion in the x direction

$$\frac{1}{\hbar} \int p_x dx + \gamma = 2\pi n . \tag{16}$$

The integral is over one period of the motion, and γ is the sum of the phase shifts acquired at the two classical turning points[3]. The phase shift upon reflection at the boundary is π, as in optics upon reflection of a metallic mirror. The other turning point is a *caustic*, giving rise to a phase shift[4] of $-\pi/2$. Using (14–16) we can thus write

$$\frac{eB}{m} \int (Y-y) dx = 2\pi(n - \tfrac{1}{4}), \qquad n = 1,2,... \tag{17}$$

This quantization rule has the simple geometrical interpretation that the flux enclosed by one arc of the skipping orbit and the boundary equals $(n - \tfrac{1}{4})$ times the flux quantum h/e (see Fig.6). This implies that the angle α_n, under which the skipping orbit is reflected from the boundary, is quantized. Simple geometry shows that

$$\frac{\pi}{2} - \alpha_n - \tfrac{1}{2}\sin 2\alpha_n = \frac{2\pi}{k_F l_{\text{cycl}}} (n - \tfrac{1}{4}), \qquad n = 1,2,...N , \tag{18}$$

with N the largest integer smaller than $\tfrac{1}{2}k_F l_{\text{cycl}} + \tfrac{1}{4}$. (For simplicity, we approximate $N \approx \tfrac{1}{2}k_F l_{\text{cycl}}$ in the rest of this paper). It follows from Fig.6 that $k_n = k_F \sin\alpha_n$. The dependence of the wavenumber of each edge state on the mode index n is thus implicitly contained in (18).

7.3 Mode–interference and Coherent Electron Focusing

In optics, a coherent point source results if a small hole in a screen is illuminated by an extended source. Such a point source excites the states on the other side of the screen coherently. A small point contact acts similarly, at least in weak magnetic fields, which is the reason why we can perform coherent electron focusing experiments. The wave function Ψ resulting from the coherent excitation at $y = 0$

---

[3]An essential difference between light optics and electron optics is hidden in this expression. Wave fronts are perpendicular to light rays. This is not the case for electrons in a magnetic field, because the phase velocity (in the direction of the canonical momentum $\mathbf{p}$) is not parallel to the electron velocity $\mathbf{v}$. Due to this "skew connection" [27] the phase difference between paths connecting two points is determined by the sum of the optical path length difference and the Aharonov–Bohm phase coresponding to the magnetic flux piercing the area enclosed. This phase is included via the canonial momentum in (16), and has to be taken into account explicitly in a trajectory treatment of coherent electron focusing [9,10].

[4]A lens cannot focus a particle flux tube to a point, because diffraction sets a lower limit to the flux tube cross section [37]. At a small separation R from a caustic the cross section is proportional to R, so that the amplitude $A \propto R^{-1/2}$. The sign change of R upon passing through the caustic then leads to the phase factor $(-1)^{-1/2} = \exp(-i\pi/2)$.

is the form

$$\Psi(x,y) = \sum_{n=1}^{N} w_n f_n(x)\exp(ik_n y). \tag{19}$$

Here $f_n(x)$ is the transverse wave function of mode n, and w_n its excitation factor. The collector voltage is in a first approximation (in the regime $W << \lambda_F$), determined by the probability density of the wave function at an infinitesimal distance from the boundary, unperturbed by the presence of the collector point contact. In the coherent electron focusing regime, the probability density near the collector at $y = L$ is dominated by the phase factors $\exp(ik_n L)$, which vary rapidly as a function of n. The weight factors w_n depend on the modeling of the point contacts, but do not affect the qualitative features of the focusing spectrum.

For point contacts modeled as an ideal point source, the transmission probability from injector to collector $T_{i\ c}$ occurring in the general formula (13) can be written as a product of three sequential transmission probabilities[5]:
1. Through the injector point contact with probability $G_i/(2e^2/h)$.
2. From the vicinity of the injector at $(x,y) = (0,0)$ to a point $(0,L)$ near the collector point contact. This is directly proportional to the probability density close to the collector, which can be expressed as the square of a sum of N plane waves (cf. (19)).
3. From there through the collector with probability $G_c/(2e^2/h)$. Using a weight factor derived in Ref. 9, we thus find

$$\frac{V_c}{I_i} = \frac{h}{2e^2} \left| \frac{1}{N} \sum_{n=1}^{N} e^{ik_n L} \right|^2 \tag{20a}$$

$$= \frac{h}{2e^2} \frac{1}{N^2} \left[N + \sum_{n \neq n'} \sum_{n'} \cos[(k_n - k_{n'})L] \right], \tag{20b}$$

The first term in (20b) is the incoherent contribution, which gives $V_c/I_i = (h/2e^2)N^{-1}$. This is just the ordinary Hall effect, without corrections due to focusing. The second term in (20b) represents a sum of oscillations in the transmission probability, expressed in terms of interference between pairs of edge states with different wave numbers.

In the experiment B is varied, affecting the values of k_n and N. A plot of a numerical evaluation of (20) (for the experimental parameters of Fig.5) is shown in Fig.7. As in the experiment, we find fine structure[6] on classical focusing peaks at

---

[5]This is in contrast to the opposite limit of adiabatic transport, realized for wider point contacts in strong magnetic fields [38,9].

[6]We have not attempted to model the partial spatial coherence due to the finite point contact size. This would introduce a coherence length into the problem, reflecting the properties of the point contact source, to be distinguished from the coherence length related to inelastic scattering, known in the field of transport in disordered conductors [12,13,18], which is a property of the medium.

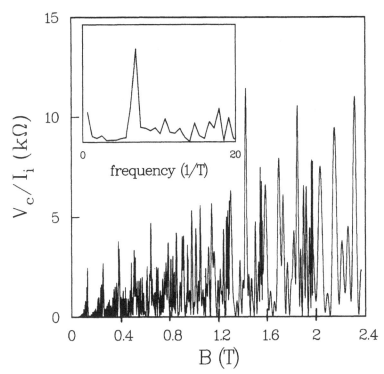

Fig.7   Theoretical electron focusing spectrum calculated from (20) for $L = 1.5$
         μm. Inset shows the Fourier power spectrum for $B > 0.8$ T. No
         correction has been made for the finite width of the point contacts in the
         experiment of Fig.5.

low magnetic fields, which becomes entirely dominant at higher fields. It is
apparent from Fig.7 (and confirmed by Fourier transform) that the
*large–amplitude high–field oscillations have the same periodicity as the smaller
low–field peaks* –as observed experimentally. This is the main result of our
calculation, which we have found to be insensitive to details of the point contact
modeling. (We have checked numerically that contributions due to evanescent
waves, neglected in (20), are small). The calculation presented here assumes that
all edge states are excited by the point contact (see Ref.9). The case of selective
excitation of edge states which results if the point contacts are flared into a horn is
discussed elsewhere [11].
If all N modes arrive in phase at the collector, (20) yields the *upper bound* of the
focusing signal $V_c/I_i = h/2e^2 \approx 12.9k\Omega$, enhanced by a factor N over the incoherent
result. The minimum intensity due to destructive interference is zero. The largest
oscillations observed experimentally (see Fig.5) are from 0.3 to 10 kΩ, which is
close to these limits. Peaks of comparable magnitude are seen in the calculated
spectrum (Fig. 7). This demonstrates that a nearly ideal coherence between the
different edge states has been realized in this experiment.
     We now turn to a qualitative discussion of the origin of the periodicity in the
focusing spectrum. The dependence on n of the phase $k_n L$ at the collector is close

to linear in a broad interval. Expansion of (18) around $\alpha_n = 0$ gives

$$k_n L = \text{constant} - 2\pi \frac{B}{B_{\text{focus}}} + k_F L \times \text{order} \left[\frac{N-2n}{N} \right]^3 . \tag{21}$$

Here $B_{\text{focus}} \equiv 2\hbar k_F/eL$ is the same as the magnetic field which follows from the classical focusing condition $2l_{\text{cycl}} = L$ discussed in Sec. 5. It follows from this expansion that, if B/B_{focus} is an integer, a fraction of order $(1/k_F L)^{1/3}$ of the N edge states interfere constructively at the collector. Because of the 1/3 power this is a substantial fraction, even for the large $k_F L \approx 225$ of the experiment. This is confirmed by the plot in Fig.8 of the phase for each edge state (modulo 2π) obtained from a numerical solution of (18), as a function of n for the second focusing peak. We conclude that the enhanced magnitude of the high field peaks with the classical focusing periodicity is a consequence of the constructive interference of a large fraction of the coherently excited edge states. The classical focusing spectrum[7] is regained in the limit $\lambda_F/L \to 0$. Raising the temperature induces a smearing of the focusing spectrum, but the line shape does not approach the classical one [9].

The edge states outside the domain of linear n–dependence of the phase give rise to additional interference structure, which does not have a simple periodicity. From (20b), it is clear that the oscillations are, generally, determined by the *mode interference wavelengths*

$$\Lambda_{n,n'} \equiv 2\pi/(k_n - k_{n'}) . \tag{22}$$

Two modes n and n' arrive in phase at the collector, and thus interfere constructively, if the distance between injector and collector L is an integer multiple of their mode–interference wavelength. The wave number k_n has values between $\pm k_F$. It follows from (18) that the *largest* mode interference wavelength $\Lambda_{\text{max}} = 2l_{\text{cycl}}$ (corresponding to interference between mode $n = N/2$ with its nearest neighbors[8]). This wavelength is associated with the dominant periodic oscillations, discussed above. The *shortest* mode interference wavelength is $\Lambda_{\text{min}} = \pi/k_F$ (corresponding to interference between mode $n = 0$ and $n = N$). this is shorter by a factor $\pi/2k_F l_{\text{cycl}} \approx 1/N$, the number of edge states. This explains why the fast oscillations in the focusing spectrum disappear at strong fields, where only a few edge states are occupied. We remark that in the latter case the argument based on (21) breaks down, and as witnessed by the magnitude of some of the

---

[7]To be more precise, in the limit $\lambda_F/L \to 0$, $\lambda_F L = \text{constant}$, one obtains focusing peaks with the classical B_{focus} periodicity, and with negligible fine structure. However, the shape and height of the peaks will be different from the classical result in Fig. 3a, if one retains the condition that the point contact width $W \lesssim \lambda_F$. For a fully classical focusing spectrum one needs also that $\lambda_F/W \to 0$.

[8]Using $k_n = k_F \sin \alpha_n$ and differentiating (18), we find a general expression for the mode–interference wavelength for neighboring modes $\Lambda_{n,n-1} = 2(l_m)^2\{(k_F)^2 - (k_n)^2\}^{1/2}$ which is recognized as the chord length of the skipping orbit corresponding to the edge state n (see Fig.6).

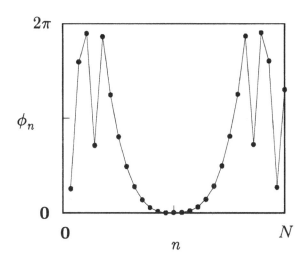

Fig.8   Phase $\phi_n = k_n L$ (modulo 2π) of each edge state at the collector for
          $L/2l_{\text{cycl}} = 2$, and $N = k_F l_{\text{cycl}}/2 = 28$. Edge states with n in the
          neighborhood of $N/2$ interfere constructively.

focusing peaks in Fig. 7 (or in the experiment), occasionally *all* the modes interfere
constructively.

   Our theory accounts for the most important novel features of the experiment,
which are not observed in the classical ballistic transport regime in metals. A more
quantitative comparison between theory and experiment requires a detailed
analysis of the potential at the point contact and at the 2DEG boundary,
something which we have not attempted. Also, we surmise that an exact treatment
of the focusing for infinitesimally narrow point contacts on an exact infinitely
repulsive potential boundary would be possible, since the transmission probabilities
can then be expressed in terms of the unperturbed wave functions of the edge
states –which are known exactly (Weber functions). Undoubtedly, such a
calculation would give results different in detail from the calculated spectrum in
Fig.7. It is most likely, however, that –as in the case of related phenomena
mentioned below– the uncertainties in the experimental conditions [9] will preclude
a better agreement with an exact theory. We stress that the observed appearance
of high–field oscillations with the focusing periodicity, but with much large
amplitude is characteristic for the mode–interference mechanism proposed.

   Coherent electron focusing is a nice demonstration of a transmission experiment
in the quantum ballistic transport regime. It has also yielded information of a more
specific nature [8–10]: 1. Boundary scattering in a 2DEG is highly specular,
capable of sustaining descriptions of ballistic transport in terms of skipping orbits
and magnetic edge states. 2. Electron motion in the 2DEG is ballistic and coherent
over distances of several microns. 3. A small quantum point contact acts as a
monochromatic point source, exciting a coherent superposition of edge states. By
combining these properties in the electron focusing geometry, a quantum
interference effect leading to a conductance modulation of nearly 100% is realized.

8. OTHER MODE–INTERFERENCE PHENOMENA

In this paper we have discussed coherent electron focusing as a manifestation of electron optics in the solid state. Such analogies provide us with valuable insight if used with caution, and they can stimulate new experiments. We mention some other interesting analogies taken from different fields. Very long wavelength radio waves (λ of the order of 1 km) propagate around the earth as in a waveguide, bounded by the parallel curved conducting surfaces formed by the earth and the ionosphere. The guiding action explains why Marconi in 1902 could be successful in transmitting radio signals across the Atlantic ocean [39]. Mode–interference is commonly observed as fading in radio signals at sunrise or sunset. Focusing of guided sound waves occurs in the ocean as a result of a vertical refractive index profile (due to gradients in hydrostatic pressure, salinity and temperature). The mode– and ray–treatments of the resulting interference patterns in the acoustic pressure have many similarities with our theory of coherent electron focusing [40,41]. The propagation of curved rays along a plane surface (as in electron focusing), is formally equivalent to that of straight rays along a curved surface. We mention in this connection the clinging of sound to a curved wall, which is the mechanism responsible for the "whispering gallery" effect in St. Paul's cathedral, explained by Lord Rayleigh [42], and for the "talking wall" in the Temple of Heaven in Peking [41].

Quantum ballistic transport can be studied in many geometries, and a wealth of results has been obtained by various groups [12–14]. Many aspects of the present discussion have an applicability beyond the specific context of electron focusing. For example, the description of transport in terms of excitation, detection and interference of quantum subbands or magnetic edge states as modes in a transmission problem, is equally significant for the Aharonov–Bohm effect in small 2DEG rings or for transport in multi–probe "electron waveguides" [43–47]. These concepts have also been applied to the quantum Hall effect [9,38,48,49]. Another area of research concerns the reproducible conductance fluctuations observed as a function of magnetic field or gate–voltage in small disordered electronic systems. This field has been extensively explored in the diffusive transport regime, and a standard theory of these "universal conductance fluctuations" in terms of quantum interference of the conduction electrons on random paths is available [13,50]. The physics underlying these fluctuations changes, however, as the ballistic transport regime is approached, because of the increasing importance of boundary scattering [17,51]. For channels with a width comparable to the Fermi wavelength, a description of transport in terms of subbands, or modes becomes the natural one. A theory for fluctuations in this regime is not yet available. We believe that the concept of mode interference, discussed here for the purely ballistic transport regime, may present a useful point of departure. Some recent theoretical work proceeds in this direction [52,53], while several transport measurements may already be in the relevant regime [54].

ACKNOWLEDGMENT

The authors have greatly benefitted from their collaboration with M.E.I. Broekaart, C.T. Foxon, J.J. Harris, L.P. Kouwenhoven, P.H.M. van Loosdrecht, D. van der Marel, J.E. Mooij, J.A. Pals, M.F.H. Schuurmans, B.J. van Wees and J.G. Williamson. Valuable discussions with D. van der Marel on the optical analogue of the conductance quantization are gratefully acknowledged.

·

REFERENCES

[1] O. Klemperer and M.E. Barnett, *Electron Optics* (Cambridge University Press, Cambridge) (1971).
[2] H. Busch, Ann. Phys. Lpz. 81 (1926) 974.
[3] Yu.V. Sharvin, Zh.Eksp.Teor.Fiz. 48 (1965) 984 [Sov.Phys. JETP 21 (1965) 655].
[4] V.S. Tsoi, Pis'ma Zh.Exp.Teor.Fiz. 19 (1974) 114, [JETP Lett. 19 (1974) 70].
[5] T. Ando, A.B. Fowler and F. Stern, Rev. Mod. Phys. 54 (1982) 437; *The Quantum Hall Effect*, R.E. Prange and S.M. Girvin, eds. (Springer, New York, 1987).
[6] B.J. van Wees, H. van Houten, C.W.J. Beenakker, J.G. Williamson, L.P. Kouwenhoven, D. van der Marel and C.T. Foxon, Phys. Rev. Lett. 60 (1988) 848; B.J. van Wees *et al.* Phys. Rev. B38 (1988) 3625.
[7] D.A. Wharam, T.J. Thornton, R. Newbury, M. Pepper, H. Ahmed, J.E.F. Frost, D.G. Hasko, D.C. Peacock, D.A. Ritchie and G.A.C. Jones, J.Phys. C21, (1988) L209.
[8] H. van Houten, B.J. van Wees, J.E. Mooij, C.W.J. Beenakker, J.G. Williamson and C.T. Foxon, Europhys. Lett. 5 (1988) 721.
[9] H. van Houten, C.W.J. Beenakker, J.G. Williamson, M.E.I. Broekaart, P.H.M. Loosrecht, B.J. van Wees, J.E. Mooij, C.T. Foxon and J.J. Harris, Phys. Rev. B39 (1989) 8556.
[10] C.W.J. Beenakker, H. van Houten and B.J. van Wees, Europhys. Lett. 7 (1988) 359.
[11] C.W.J. Beenakker, H. van Houten and B.J. van Wees, *Festkörperprobleme/Advances in Solid State Physics* Vol. 29, ed. U. Rössler (Pergamon/Vieweg, Braunschweig, 1989) 299.
[12] H. Heinrich, G. Bauer and F. Kuchar, eds., *Physics and Technology of Submicron Structures* (Springer–Verlag, Berlin, 1988).
[13] P.A. Lee, R.A. Webb and B.L. Altshuler, eds. *Mesoscopic Phenomena in Solids* (Elsevier Science Pub., Amsterdam, 1989) to be published.
[14] M. Reed and W.P. Kirk, eds., *Nanostructure Physics and Fabrication*, (Academic Press, New York, 1989) to be published.
[15] K. von Klitzing, G. Dorda and M. Pepper, Phys. Rev. Lett. 45 (1980) 494.
[16] C.W.J. Beenakker, H. van Houten and B.J. van Wees, Superlattices and Microstructures 5 (1989) 127.
[17] H. van Houten, B.J. van Wees and C.W.J. Beenakker, in Ref. [12]
[18] Y. Imry, *Directions in Condensed Matter Physics* Vol. 1, eds. G. Grinstein and G. Mazenko (World Scientific, Singapore, 1986) 102.
[19] L.I. Glazman, G.B. Lesovick, D.E. Khmelnitskii, R.E. Shekhter, Pis'ma Zh. Teor. Fiz, 48 (1988) 218 [JETP Lett. 48 (1988) 238].
[20] R. Landauer, Z.Phys.B. 68 (1987) 217.
[21] C.W.J. Beenakker, H. van Houten, Phys. Rev. B39 (1989) 10445; H. van Houten, C.W.J. Beenakker, in Ref. 14.
[22] E.G. Haanappel and D. van der Marel, Phys. Rev. B39 (1989) 5484; D. van der Marel and E.G. Haanappel, *ibid*, B39 (1989) 7811.
[23] A. Szafer and A.D. Stone, Phys. Rev. Lett. 62 (1989) 300.
[24] G. Kirczenow, Solid State Comm. 68 (1988) 715.

[25] I.B. Levinson, Pis'ma Zh.Eksp.Teor.Fiz, 48 (1988) 273 [JETP Lett. 48 (1988) 301].
[26] Song He and S. Das Sarma, Phys. Rev. B40 (1989) 3379; E. Tekman and S. Ciraci, Phys. Rev.B39 (1982) 8772; A. Kawabata, J. Phys. Soc. Japan 58 (1989) 372.
[27] M. Born and E. Wolf, *Principles of Optics*, 3^d rev. ed. (Pergamon, Oxford, 1965).
[28] P.A.M. Benistant, Ph.D Thesis, University of Nijmegen, The Netherlands (1984); P.A.M. Benistant, A.P. van Gelder, H. van Kempen and P. Wyder, Phys. Rev. B32, 3351.
[29] M. Büttiker, Phys. Rev. Lett. 57 (1986) 1761; IBM J. Res. Dev. 32 (1988) 317.
[30] G.C. Southworth, *Principles and Applications of Waveguide Transmission* (Van Nostrand, Toronto, 1950) 25.
[31] R. Landauer, IBM J. Res. Dev. 1 (1957) 223; see also Ref.20.
[32] D.S. Fisher and P.A. Lee, Phys. Rev. B23 (1981) 6851.
[33] A.D. Stone and A. Szafer, I.B.M. Journal of Res. and Dev. 32 (1988) 384; H.V. Baranger and A.D. Stone, Phys. Rev. Lett. 63 (1989) 414.
[34] J. Kucera and P. Streda, J. Phys. C21 (1988) 4357.
[35] R.E. Prange and T.-W. Nee, Phys. Rev. 168 (1968) 779.
[36] M.S. Khaikin, Adv. Phys. 18 (1969) 1.
[37] L.D. Landau and E.M. Lifshitz, *The Classical Theory of Fields*, 4th ed. (Pergamon, Oxford, 1987) par. 54.
[38] B.J. van Wees, E.M. Willems, C.J.P.M. Harmans, C.W.J. Beenakker, H. van Houten, J.G. Williamson, C.T. Foxon and J.J. Harris, Phys. Rev. Lett. 62 (1989) 1181.
[39] K.G. Budden, *The Waveguide Mode Theory of Wave Propagation* (Prentice–Hall, London, 1961).
[40] I. Tolstoy, *Proc. Symp. on Quasi–Optics*, Microwave Research Institute Symposia Series Vol. XIV (Polytechnic Press, New York, 1964) 43.
[41] L.M. Brekhovskikh, *Waves in Layered Media* (Academic Press, New York, 1960).
[42] Lord Rayleigh, *Theory of Sound*, 2nd ed. (Macmillan, London, 1896) par. 287; Phil. Mag. 20 (1910) 1001; Phil. Mag. 27 (1914) 100.
[43] G. Timp, A.M. Chang, P. Mankiewich, R. Behringer, J.E. Cunningham, T.Y. Chang and R.E. Howard, Phys. Rev. Lett. 59 (1987) 732; G. Timp in Ref. 13; G. Timp, P.M. Mankiewich, P. de Vegvar, J.E. Cunningham, R.E. Howard, H.V. Baranger and J.K. Jain, Phys. Rev. B39 (1989) 6227.
[44] A.M. Chang, K. Owusu–Sekyere and T.Y. Chang, T.Y., Solid State Com. 67 (1988) 1027.
[45] C.J.B. Ford, T.J. Thornton, R. Newbury, M. Pepper, H. Ahmed, D.C. Peacock, D.A. Ritchie, J.E.F. Frost and G.A.C. Jones, Phys. Rev. B 38 (1988) 8518; T.J. Thornton, *et al.* in Ref. 12.
[46] M.L. Roukes, A. Scherer, S.J. Allen Jr., H.G. Graighead, R.M. Ruthen, E.D. Beebe and J.P. Harbison, Phys. Rev. Lett. 59 (1987) 3011.
[47] K. Ishibashi, Y. Takagaki, K. Gamo, S. Namba, S. Ishida, K. Murase, Y. Aoyagi and M. Kawabe Solid State Comm. 64 (1987) 573.
[48] M. Büttiker, Phys. Rev. B., 38 (1988) 9375.
[49] H. van Houten, C.W.J. Beenakker, P.H.M. van Loosdrecht, T.J. Thornton, H. Ahmed, M. Pepper, C.T. Foxon and J.J. Harris, Phys. Rev.B37 (1988) 8534; R.J. Haug, A.H. MacDonald, P. Streda and K. von Klitzing, Phys.

Rev. Lett., 61 (1988) 2797; S. Washburn, A.B. Fowler, H. Schmidt and D. Kern, Phys. Rev. Lett. 61 (1988) 2801.

[50] P.A. Lee, A.D. Stone and H. Fukuyama, Phys. Rev. B35 (1987) 1039.

[51] C.W.J. Beenakker and H. van Houten, Phys. Rev. B37 (1988) 6544.

[52] Z. Tesanovic, M.V. Jaric and S. Maekawa, Phys. Rev. Lett. 57 (1986) 2760; Y. Isawa, Surf. Sci. 170 (1986) 38.

[53] X.C Xie and S. Das Sarma, Solid State Comm. 68 (1988) 697.

[54] J.H.F. Scott–Thomas, St.B. Field, M.A. Kastner, H.I. Smith and D.A. Antoniadis, Phys. Rev.Lett. 62 (1989) 583; A.M. Chang, G. Timp, R.E. Howard, R.E. Behringer, P.M. Mankiewich, J.E. Cunningham, T.Y. Chang and B. Chelluri, Supperlattices and Microstructures, 4 (1988) 515.

H. van Houten is with Philips Research Briarcliff, NY 10510, USA; his present address is Philips Research Laboratories, 5600 JA Eindhoven, The Netherlands, *C.W.J. Beenakker* is with Philips Research Laboratories, 5600 JA Eindhoven, The Netherlands.

CHARGE BUILD-UP AND INTRINSIC BISTABILITY IN RESONANT TUNNELING

L. Eaves

This chapter reviews the effect of space charge build–up in the quantum well of resonant tunneling devices. Space charge build–up can be particularly important in devices with thick collector barriers and requires a modification of the viewpoint of resonant tunneling as an electrical analogue of the Fabry–Perot interferometer. In particular, space charge build–up leads to a novel hysteresis effect in the current–voltage characteristics which is termed intrinsic bistability. Some of the developments leading to the experimental realisation of intrinsic bistability and of our theoretical understanding of it are outlined. The major part of the chapter describes how measurements of the tunnel current and capacitance in a quantising magnetic field provide quantitative information about the charge build–up process and the energy relaxation of carriers in the quantum well.

1. INTRODUCTION

It is now sixteen years since Chang, Esaki and Tsu [1] reported the first observation of resonant tunneling in a double barrier heterostructure device. This device can be thought of as the electronic analogue of a Fabry–Perot interferometer in which the quantum well, sandwiched between two tunnel barriers, acts as a resonant cavity. The applied voltage is used to "tune" the de Broglie wavelength of the electrons injected into the cavity from the negatively–biased emitter contact. The resonances are observed as peaks in the current voltage characteristics $I(V)$ with regions of negative differential conductivity (NDC) beyond the peaks. Much of the work on resonant tunneling still concentrates on the lattice–matched heterostructure system based on $GaAs/Al_xGa_{1-x}As$. However, other III–V semiconductor systems can also be prepared with very high quality, including those based on pseudomorphic strain layers (Hiyamizu *et al.* [2]). The resonant tunneling device forms the heart of several recently developed uni– and bi–polar devices (for reviews see Capasso [3], Luryi [4]). As a result of the improvement of material quality, electrons can travel ballistically over long distances in the quantum well and resonant tunneling and novel magneto–electric quantisation effects have been observed in devices with well widths up to 180 nm (Snell *et al.* [5]; Eaves *et al.* [6]; Alves *et al.* [7]; England *et al.* [8]; Henini *et al.* [9]; Leadbeater *et al.* [10]).

Considerable interest and controversy has been generated by the argument about whether resonant tunneling should be regarded as a coherent or sequential process (Payne [11]; Weil and Vinter [12]; Luryi [13]). In the sequential model,

W. van Haeringen and D. Lenstra (eds.), Analogies in Optics and Micro Electronics, 227–242.
© 1990 *Kluwer Academic Publishers. Printed in the Netherlands.*

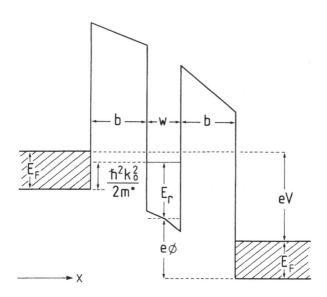

Fig. 1   Spatial variation of electron potential energy through an idealised
           double barrier structure as modeled by Sheard and Toombs, showing
           bound state and Fermi levels.

resonantly tunneling electrons undergo scattering processes in the quantum well,
thereby losing quantum mechanical phase coherence before tunneling out through
the collector barrier. The effect of scattering processes on the tunneling electrons,
particularly those arising from the emission of longitudinal optical phonons, reveal
themselves clearly in the current voltage characteristics in the presence of a
quantising magnetic field (Leadbeater et al. [14]).

    When the second barrier has a relatively low transmission coefficient, a
significant electronic space–charge can build up in the well at resonance (Ricco and
Azbel [15]). This charge must be taken into account when considering the electrical
properties of resonant tunneling devices and gives rise to intrinsic bistability in the
current–voltage characteristics, $I(V)$. Since this effect was first proposed by
Goldman et al. [16]), it has been the subject of some controversy (see Sollner [17],
Goldman et al. [18], Toombs et al. [19], Leadbeater et al. [14], Foster et al. [20]).
However, intrinsic bistability has recently been observed definitively (Alves et al.
[21], Zaslavsky et al. [22], Leadbeater et al. [23]) and forms the major topic of this
chapter.

2. INTRINSIC BISTABILITY

Sheard and Toombs have considered the problem of bistability using the sequential
tunneling approach [24] in which transmission of electrons is regarded as two
successive transitions, from emitter contact into the bound state of the well and
then from the well into the collector contact. The simplified model of a symmetric

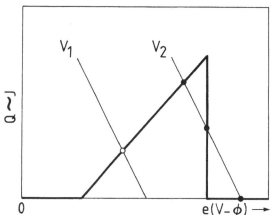

Fig. 2   Schematic diagram showing how intrinsic bistability arises in the
Sheard–Toombs model. Q is plotted against voltage drop across emitter
barrier for (a) equation obtained from dynamics of electrons entering
and leaving well (thick lines) and (b) equation governing the
electrostatics (thin lines).

double barrier structure (DBS) is shown in Fig.1 under an applied voltage V. Here
$E_f = \hbar^2 k_f^2/2m^*$ is the Fermi energy of electrons in the emitter (left hand) and
collector (right hand) contacts and m^* is the electron effective mass ($m^* = 0.07m_e$
for GaAs). The energy of the quasi–bound state in the well is E_r and w and b are
the well and barrier thicknesses respectively. The dynamics of electrons entering
and leaving the well and the effect (electrostatic feedback) of charge build–up in
the well on the distribution of voltage between the emitter and collector barriers
provides two equations relating the charge Q in the well to the applied voltage V.
The dependence of Q on V is determined by the simultaneous solutions of these
equations, as shown schematically in Fig.2. For simplicity, the transmission
coefficients T_e and T_c of the emitter and collector barrier are treated as constants,
independent of V. The intersection points of the two plots give the required
solution $Q(V)$. Since the current density $J = Q/\tau_c$, where $1/\tau_c$ is the decay rate of
stored charge, the idealised current–voltage curves shown in Fig.3 can be obtained
from $Q(V)$. The threshold voltage $V_{th} = 2(E_r - E_f)/e$ and the feedback parameter
$\alpha = (2b + w + 2\lambda_s)T_e/a_0(T_e + T_c)$, where a_0 is the effective Bohr radius ($a_0 = 10$
nm for GaAs) and λ_s is the screening length of the n$^+$ electrons (assumed
constant). Note also that above threshold, the current increases lineary with V
since $J = Q_m e(V - V_{th})/2E_f(1 + \alpha)\tau_c$, where Q_m is the charge density in the well
at peak current. The dashed lines in Fig.3 indicate the transitions between
high–charge and zero–charge states which would occur on sweeping the voltage up
and down through the region of resonant tunneling. In the sequential model, the
time required for such a transition is τ_c. The voltage width of the region of
intrinsic bistability is $2\alpha E_f = Q_m/C$, where $C = \epsilon_r \epsilon_0/(b + \tfrac{1}{2}w + \lambda_s)$. It is
interesting to note the similarity of the solutions given by Figs. 2 and 3 to the
well–known load–line problem involving a series resistance R_l coupled to a device
exhibiting negative differential conductivity. When the two series elements

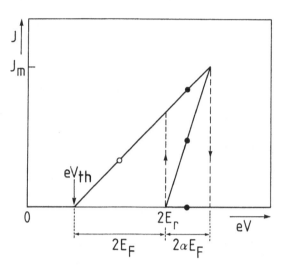

Fig. 3   Current–voltage characteristic J versus eV, showing region of
           bistability. The open and solid circles refer to the intersection points of
           Fig.2.

are biased in the region of NDC, a bistable DC current (no oscillations) is obtained
when $R_1 |dI/dV|_{ndc} > 1$.

Goldman *et al.* [16] were the first to propose and search for intrinsic bistability
in DBS. Studies of the current–voltage characteristics as a function of magnetic
field by Goldman *et al.* [25] and Payling *et al.* [26,27] have clearly demonstrated
charge build–up in the well above the threshold voltage. However, Sollner [17]
argued that the bistability Goldman *et al.* observed in the $I(V)$ characteristics was
not an intrinsic property of the DBS itself, but due to oscillations of the current in
the circuit containing the DBS when biased in the region of negative differential
conductivity (NDC). Indeed, it is worth noting that similar bistabilities can also be
observed in other devices (e.g. Esaki and Gunn diodes) which exhibit NDC. This
"extrinsic" bistability in $I(V)$ gives rise to a characteristic double–stepped decrease
in I around the region of NDC, very similar to that originally observed by
Goldman *et al.* [16]. It arises because the current through the DBS and its circuit
breaks into oscillation at different voltages on the up– and down–voltage sweeps.
By solving numerically the differential equation of a circuit comprising a device
exhibiting NDC (but no intrinsic bistability) together with passive R, L and C
elements, Toombs *et al.* [19] show that "extrinsic" bistability effects very similar
to those reported by Goldman *et al.* could be simulated. Note that this bistability
is quite different from the current–related DC bistability (no oscillations) which
occurs when a large series resistance is connected to an element with NDC. The
extrinsic bistability and related current oscillations can be eliminated by
connecting a small chip resistor ($r < |dV/dI|_{NDC}$) in parallel with the DBS using
very short connecting leads. A stable $I(V)$ plot of the combined device and parallel
resistance can then be obtained by applying a ramped voltage. The $I(V)$
characteristics of the device alone are calculated by subtracting, from the total

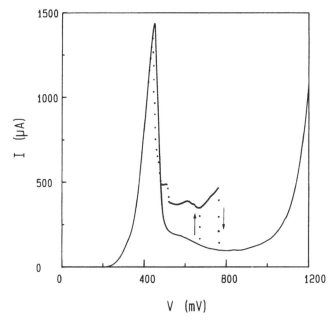

Fig. 4   The current–voltage characteristics at 77 K of a 5 μm diameter mesa
            fabricated from the double–barrier structure I (5 nm well). The dotted
            curve was measured when the device was oscillating and the full curve
            when the oscillation and current bistability were suppressed by
            connecting a small resistor (25 Ω) in parallel with the device.

current, the current V/r flowing through the resistor. Typical results are shown in
Fig.4 for a 5 μm diameter mesa DBS fabricated from a structure (structure I)
comprising the following layers in order of growth from the n^+ substrate: (i) 1.0 μm
of $n = 2\times10^{18}$ cm^{-3} GaAs; (ii) 50 nm of $n = 2\times10^{16}$ cm^{-3} GaAs; (iii) 2.5 nm of
undoped GaAs; (iv) 5.6 nm of undoped (AlGa)As, [Al] = 0.4; (v) 5 nm of undoped
GaAs; (vi) 5.6 nm of undoped (AlGa)As, [Al] = 0.4; (vii) 2.5 nm of undoped GaAs;
(viii) 50 nm of $n = 2\times10^{16}$ cm^{-3} GaAs and (ix) 0.5 μm of $n = 2\times10^{18}$ cm^{-3} GaAs. No
bistability effect is observed in $I(V)$ when the parallel resistor is used. The voltage
range between threshold and peak is narrow, indicating that the charge build–up
and electrostatic feedback is small. In this case the "voltage overhang" $2\alpha E_F$ (see
Fig.3) required for bistability is also small. Hence the intrinsic bistability effect
might be washed out by broadening mechanisms, e.g. atomic monolayer
fluctuations in the well thickness.
      In order to enhance the electrostatic feedback we have prepared and
investigated a structure (II) in which the two barriers have different widths. It
comprises the following layers, in order of growth from the n^+ substrate: (i) 2 μm of
$n = 2\times10^{18}$ cm^{-3} GaAs; (ii) 50 nm of $n = 1\times10^{17}$ cm^{-3} GaAs; (iii) 50 nm of $n =$
1×10^{16} cm^{-3} GaAs; (iv) 3.3 nm of undoped GaAs; (v) 8.3 nm of undoped (AlGa)As,
[Al] = 0.4; (vi) 5.8 nm of undoped GaAs; (vii) 11.1 nm of undoped (AlGa)As, [Al]
= 0.4; (viii) 3.3 nm of undoped GaAs; (ix) 50 nm of $n = 1\times10^{16}$ cm^{-3} GaAs; (x) 50

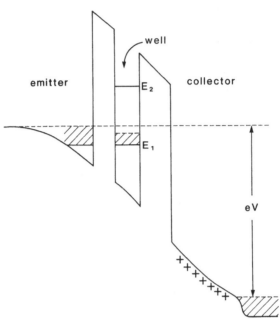

Fig. 5   Conduction–band profile of the asymmetric double–barrier structure
under applied voltage V showing bound–state levels (solid lines) in the
accumulation layer of the emitter and in the well. The quasi–Fermi
levels in the emitter, well and collector are also shown (dashed lines).

nm of $n = 1{\times}10^{17}$ cm^{-3} GaAs; and (xi) 0.5 μm of $2{\times}10^{18}$ cm^{-3} GaAs top contact.

The layers were processed into mesas of diameter 200 μm. The conduction band
profile with the top contact biased positively, is shown in Fig.5. A bound state is
formed in the accumulation region adjacent to the emitter barrier and, at liquid
helium temperatures, the associated two–dimensional electron gas (2DEG) is
degenerate. The accumulation potential may be modelled satisfactorily using a
variational solution for the bound–state wave function (Fang and Howard [28]). At
a bias of 330 mV the electric field in the emitter barrier is 29 kV cm^{-1} and the
width of the accumulation region is ~14 nm. In the remainder of this layer (~36
nm) the Fermi level is close to the conduction–band edge since the doping density
(10^{16} cm^{-3}) is close to the Mott metal–insulator transition. However, this does not
give rise to an important series resistance as shown by previous studies of similarly
doped single–barrier structures (Eaves et al. [29]).

Goldman et al. [30] also investigated an asymmetric structure of this type, but
with thinner barriers. However, it showed a bistability very similar to that
characteristic of the extrinsic effect (i.e. with circuit oscillations) in both bias
directions. The $I(V)$ characteristics of our device (200 μm diameter mesa) at 4 K
are shown in Fig.6. In each bias direction two resonant peaks in $I(V)$ are observed,
corresponding to resonant tunneling into the quasi bound states (E_1 and E_2) of the
well. There is a pronounced difference between forward and reverse bias. Forward
bias is defined as substrate negative, thin emitter barrier, thick collector barrier, as
shown schematically in Fig.5. (This definition of forward bias is used for

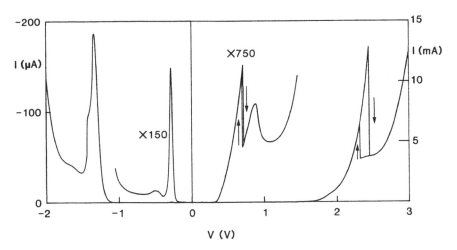

Fig. 6   The current–voltage characteristics at 4 K of a 200 μm diameter mesa fabricated from the asymmetric double–barrier structure II described in the text. Note the very strong LO phonon–assisted peak at $V = 0.85$ V.

convenience since it represents the sense for which intrinsic bistability is observed. The opposite definition was used in the original articles by Alves et al. [21] and Leadbeater et al. [23]).

In forward bias, electrons are injected from the emitter into the well through the thinner barrier. The small tunneling probability through the thicker collector barrier retains the charge in the well and enhances the electrostatic feedback. The form of the bistability observed in forward bias is quite different from that originally observed by Goldman et al. [16]. In addition, the current does not oscillate and the $I(V)$ characteristics remain unchanged when a parallel capacitor or resistor is connected. Note also that on both high and low current branches of the hysteresis loop, $dI/dV > 0$. The curve resembles a load–line effect [31] but such a similarity is expected from the theoretical analysis of Sheard and Toombs [24, 32] for the effect of electrostatic feedback on the $I(V)$ characteristics (see Fig.3). The bistabilities in the $I(V)$ characteristics cannot be explained by the presence of an ohmic series resistance.

This asymmetric DBS differs from the idealised structure shown in Fig.1. Due to the light doping in the contact layers and the large threshold voltage, a two–dimensional electron gas (2DEG) is formed in the emitter accumulation layer. The light doping also gives rise to a significant voltage drop across the collector depletion layer. This depletion layer voltage magnifies the electrostatic feedback effect. The distribution of space charge across the device and the energy of the electrons stored in the quantum well can be studied by means of magnetotransport and capacitance measurements which form the subject of the next section.

3. MAGNETOTUNNELING AND CAPACITANCE MEASUREMENTS

The application of quantising magnetic fields have proved to be invaluable for

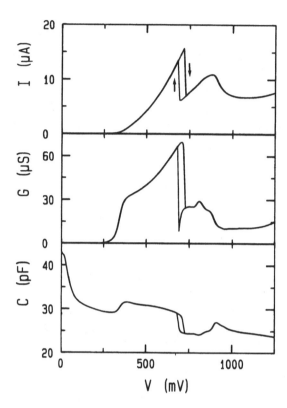

Fig. 7   Voltage dependence of DC current I, AC capacitance C and parallel
           conductance G of the asymmetric double–barrier structure measured at
           1 MHz and $T = 4$ K.

studying the electrical properties of resonant tunneling devices. In the geometry
$B \perp J$ resonant tunneling into hybrid magneto–electric states is observed (Snell *et
al.* [5], Eaves *et al.* [6], Alves *et al.* [7], England *et al.* [8], Leadbeater *et al.* [10].
Here we consider the effect of a magnetic field applied perpendicular to the barriers
$B \| J$. (For earlier work, see Mendez *et al.* [33], Mendez [34], Eaves *et al.* [6, 35].
     The current–voltage characteristic for our asymmetric DBS (structure II) at 4
K is shown in more detail in Fig.7 for the region corresponding to resonant
tunneling into the first quasi–bound state of the quantum well. The figure shows
clearly the threshold for resonant tunneling at $V_{th} = 330$ mV, the turn–off at $V_p =$
725 mV where the current peaks, and the region of intrinsic bistability. The
transverse components of momentum parallel to the barrier ($k_\|$) and hence
transverse kinetic energy ($E_\| = \hbar^2 (k_\|)^2 / 2m^*$) are conserved in the resonant
tunneling process. Total energy is also conserved so resonant tunneling occurs
when the energies of the bound states in the accumulation layer and quantum well
essentially coincide. However, because of finite level widths, the tunneling rate
from emitter into well will be a sharply peaked resonant function of the voltage

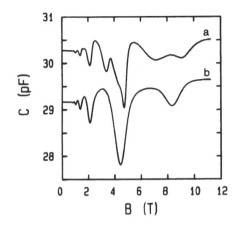

Fig. 8   Magneto–oscillations in differential capacitance C vs. magnetic field B in the asymmetric double–barrier structure for applied voltage (a) 600 mV and (b) 300 mV at $T = 4$ K and for $B\| J$.

drop across the emitter barrier. The observed resonant tunneling range V_{th} to V_p corresponds to climbing up this resonance peak. It is the electrostatic effect of the charge build–up in the quantum well (electrostatic feedback) that is responsible for the extended voltage range over which resonant tunneling is observed in the $I(V)$ curve and the appearance of current bistability. When the applied voltage is increased, only a very small voltage change across the emitter barrier is needed to charge up the well and, due to the screening effect of this charge, almost all the extra voltage drop occurs across the collector barrier and depletion layer. It is worth stressing again that when biased in the opposite direction a very sharp maximum in the $I(V)$ curve is observed (see Fig.6) since in this case (thin collector barrier) there is little charge buildup and electrostatic feedback is negligible. The bistability does not then appear.

Conservation of transverse kinetic energy in resonant tunneling requires the Fermi levels in the accumulation layer and quantum well to coincide. But the theory of resonant tunneling (Sheard and Toombs [24]) shows that the states of transverse motion in the well are only partially occupied, since the occupancy is determined dynamically by the balance between the transition rates into and out of the well. However, if the energy relaxation time is much less than the charge storage time in the well, the electron distribution will be able to thermalise to the lattice temperature and establish a well–defined quasi–Fermi level below that in the emitter contact. This is the situation illustrated in Fig.5. The Fermi energy of a thermalised 2DEG in the quantum well is less than the Fermi energy of the 2DEG in the accumulation layer since the bound–state levels essentially coincide during resonant tunneling.

This possibility can be investigated by studying magneto–oscillations in the capacitance of the structure for a magnetic field $B\| J$. In this geometry the states of transverse motion of a degenerate 2DEG (in emitter or well) are quantised into Landau levels. Theoretically, oscillations with a definite period, $\Delta(1/B)$, in $1/B$ arise from a modulation of the charge when Landau levels pass through the quasi–

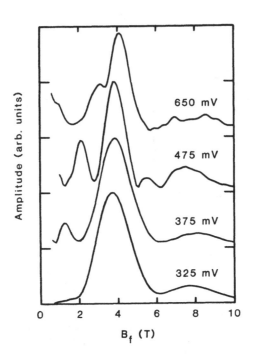

Fig. 9   Spectrum of the magneto–oscillation frequencies B_f obtained by Fourier
transforming them in $1/B$ space. The peak position B_f for each voltage
gives the inverse period. The weak structure at around 8 T corresponds
to the second harmonic of the series due to the charge in the
accumulation layer.

Fermi level. This charge modulation affects the distribution of electric
potential and screening lengths and hence modulates the capacitance of the device.
The frequency of the oscillations $B_f = \{\Delta(1/B)\}^{-1}$ is thus related to the Fermi
energy E_f by $B_f = m^*E_f/e\hbar$. For fully, as opposed to partially, occupied states
$E_f = \hbar^2\pi n/m^*$, where n is the areal electron density. This gives $n = 2eB_f/h$.
    The voltage dependence of the differential capacitance C and parallel
conductance $G = 1/R$ are plotted with that of $I(V)$ in Fig.7. The differential
parameters were measured at 1 MHz with a modulation of 3 mV using a Hewlett
Packard 4275A LCR meter in the mode which analyses the impedance of a device
as a capacitor C and parallel resistor R. Between 10 kHz and 2 MHz these
parameters are also independent of measurement frequency. Fig.8 shows typical
oscillatory structure in the variation of capacitance with magnetic field. Similar
oscillations are also observed in the current I and conductance G but are less well
defined. To reveal more clearly multiperiodic behaviour, we have Fourier analysed
the experimental capacitance traces. The distributions of magneto–oscillation
frequencies B_f thus obtained are shown in Fig.9 for different applied voltages.
Below the threshold voltage V_{th}, there is a single peak in the magneto–oscillation
spectrum. This peak is clearly associated with the 2DEG in the emitter

accumulation layer, since the corresponding areal density n_a increases steadily with voltage up to V_{th} as shown in Fig.10. Moreover, the static capacitance values, given by the ratio of total accumulation charge to applied voltage, are in good agreement with the AC values shown in Fig.7 and are consistent with our theoretical modeling of the potential distribution across the heterostructure. We have also investigated the magneto–oscillations when the magnetic field is tilted at and angle θ to the normal to the layers. Up to $\theta = 40^0$ the form of the oscillations as a function of $B\cos\theta$ is largely unchanged. This confirms the two–dimensional nature of the electrons in the emitter accumulation layer.

Between V_{th} and V_p, this magneto–oscillation frequency is independent of voltage (Figs. 9 and 10) and within experimental error, the accumulation density n_a remains constant at 2.2×10^{11} cm^{-2}. Hence the electric field and voltage drop across the emitter barrier remain virtually unchanged in this range, which is consistent with our model of resonant tunneling between bound states in the emitter and quantum well. Also in the resonant tunneling region, a second, weaker peak appears in the magneto–oscillation spectrum (Fig.9). The frequency of this peak gives an areal density n_w, which increases throughout this range and approaches n_a at the voltage V_p for maximum current (Fig.10). We attribute this peak to a degenerate electron distribution stored in the quantum well whose temperature $T_e << \hbar\omega_c/k = 15$ K at $B = 1$ T. The equality of n_a and n_w at $V = V_p$ is to be expected theoretically since the peak transition rate into the well is much greater than the decay rate out of the well throughout the collector barrier. The dynamically determined occupancy of the states in the well is then close to unity and energy relaxation has little effect on the electron distribution.

This cooling of the electrons is confirmed by comparing the lifetime τ_c of the electrons in the well with the energy relaxation rate. The lifetime τ_c is limited by tunneling through the collector barrier and is related to the current density by $J = n_w e/\tau_c$. At the resonance peak ($J \simeq 0.06$ A cm^{-2}, $n_w \simeq 2 \times 10^{11}$ cm^{-2}) this gives $\tau_c \sim 0.6$ μs. The energy relaxation must be via spontaneous emission of acoustic phonons since the temperature is low (4 K) and the electron kinetic energies ($E_f \sim 7$ meV) are too small for optic–phonon processes. An estimate of the emission rate τ^{-1} from the deformation potential gives $\tau_{ph} \sim 10^{-9}$ s for a well–width of 5.8 nm. This is indeed much shorter than τ_c. We note that electron–electron scattering is not important here since the phonon emission rate is sufficiently large to thermalise the electron distribution to the lattice temperature within the available time τ_c.

When V increases above V_p a transition occurs in which charge is expelled from the well with a consequent redistribution of potential and resonant tunneling can no longer occur. This results in a step–wise increase in the accumulation layer density to $n_a \simeq 3 \times 10^{11}$ cm^{-2}, as shown in Fig.10. For $V > V_p$ only a single magneto–oscillation period is observed. The different charge states of the device on the high– and low–current parts of the hysteresis loop are also clearly shown by the different magneto–oscillation frequencies observed. The corresponding sheet density increases smoothly with voltage showing that, in this case, there is little charge buildup in the well. We note here that the broad maximum in $I(V)$ above the current bistability is due to inelastic tunneling processes which have been previously observed and discussed elsewhere (Leadbeater et al. [14]).

The principal features of the $C(V)$ curve of Fig.7 can also be understood in terms of this model. We may regard the DBS as two parallel plate capacitors C_1 (emitter barrier and accumulation layer) and C_2 (collector barrier and depletion layer) connected in series. The charge on the common central plate corresponds to

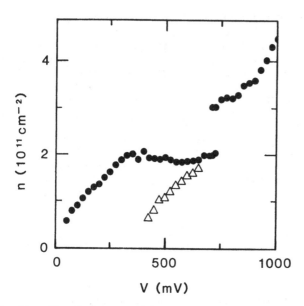

Fig. 10   Results obtained from the magneto–oscillations in the capacitance for
the asymmetric device: areal density n versus voltage V for charge in the
accumulation layer n_a (closed circles) and well n_w (triangles). The
values of n are deduced from the peaks in the Fourier spectrum.

the charge stored in the quantum well. The steep fall in capacitance at low
voltages is due to the rapid increase in depletion length in the lightly–doped (10^{16}
cm^{-3}) collector layer which decreases C_2. The slower decrease for $V > 50$ mV is
due to the less rapid rate of depletion of the layer doped to 10^{17} cm^{-3}. However, the
most notable features are the sharp increase in capacitance at V_{th} and subsequent
fall at V_p. When the bound states of the accumulation layer and quantum well are
on resonance, the voltage drop across C_1 is essentially constant. An increase in
applied voltage therefore appears almost entirely across C_2. The measured
differential capacitance is thus $C \simeq C_2$ in the resonant tunneling region, whereas
off–resonance $C = C_1 C_2/(C_1 + C_2) < C_2$.
    This quasi–static argument is not obviously applicable at 1 MHz, where the
measurements of differential capacitance were made. In a small–signal analysis
(Sheard and Toombs [32]) based on the sequential theory of resonant tunneling, C_1
and C_2 have parallel resistors R_1 and R_2 respectively. R_1 allows electrons tunneling
from the emitter to charge the quantum well whilst R_2 allows the stored charge to
leak through the collector barrier. The previous argument is equivalent to the
assumption that during resonant tunneling, R_1 becomes very small and effectively
short–circuits C_1 so that the measured capacitance $C \simeq C_2$. By identifying the time
constant $R_2 C_2$ with the storage time τ_c we have $\tau_c = R_2 C_2 \sim RC$ during resonance.
The values shown in Fig.7 give $\tau_c \sim 0.4$ μs at V_p, which is consistent with the
previous estimate from the DC current and charge. An approximate theoretical
estimate of τ_c can be made by calculating the attempt rate (obtained from the
energy of the bound state above the bottom of the well ~ 68 meV) and the

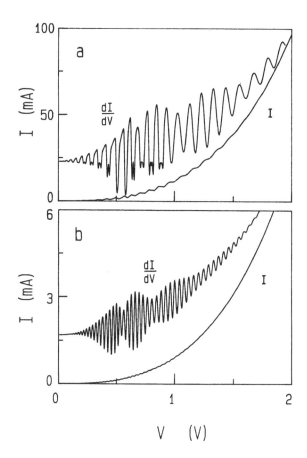

Fig. 11   Plot at $T = 4$ K of the current–voltage characteristics $I(V)$ and
differential conductance dI/dV of two double barrier structures based on
n–GaAs/Al$_{0.4}$Ga$_{0.6}$As. (a) Device with (AlGa)As barrier thickness =
5.6 nm, GaAs well width = 60 nm; (b) device with barrier thickness =
5.6 nm, well width = 120 nm.

transmission coefficient of a rectangular barrier (conduction band offset $\sim$ 320
meV). This gives $\tau_c \sim 1$ μs. Under bias a smaller value is expected since the
average collector–barrier height is then decreased. Measurement of the tunneling
escape time from the quantum well of a DBS have also been made by studying the
temporal decay of photoluminescence (Tsuchiya et al. [36]). However, the optical
determination of τ_c is limited to thin–barrier structures for which τ_c is less than
the radiative recombination time $\tau_c \sim 0.5$ ns. In our asymmetric structure it is the
thick collector barrier which give rise to the long τ_c.

The above picture of the resonant tunneling process is complemented by low
temperature photoluminescence (PL) measurements (Hayes et al. [37]). In these

measurements the sheet density in the well is determined from the linewidth of the photoluminescence spectrum produced by the recombination of photo–excited holes which tunnel into the well and recombine with resonantly tunneling electrons. The sheet density and temperature of the electrons in the well can also be deduced from the Zeeman splitting of the photoluminescence lines.

4. CONCLUSIONS

The double barrier resonant tunneling device is frequently described as the electronic analogue of the Fabry–Perot interferometer in optics. This analogy is particularly apt when the barriers are relatively thin so that space–charge build–up is small, and when the well–width is large. In this case, sharp and well–defined resonances are observed in the current–voltage characteristics. This point is illustrated in Fig.11 which plots this $I(V)$ characteristics and differential conductance of two symmetric n–type GaAs/(AlGa)As double barrier structures with $Al_{0.4}Ga_{0.6}As$ barrier thickness of 5.6 nm and well widths of 60 and 120 nm respectively. The 60 nm wide well exhibits 18 clear "Fabry–Perot–like" quantum resonances in the conductance even at room temperature. At 4 K, 28 resonances are observed. The 120 nm wide well shows over seventy resonances in the conductance at 4 K (Henini et al. [9]). Some of these correspond to resonant tunneling into states whose energy exceeds the heigth of the collector barrier (see also Sen et al. [38], Heiblum et al. [39]). However, this chapter has tried to show that electrical properties of double barrier structures are not controlled by the resonance condition alone. The electrostatic effect resulting from charge build–up in the well can strongly modify the $I(V)$ characteristics and lead to novel effects such as intrinsic bistability.

ACKNOWLEDGEMENTS

The work described in Secs. 2 and 3 has been carried out in collaboration with M.L. Leadbeater, D.G. Hayes, E.S. Alves, F.W. Sheard, G.A. Toombs, P.E. Simmonds, M.S. Skolnick, M. Henini and O.H. Hughes, and was supported by the Science and Engineering Research Council (U.K.).

REFERENCES

[1]   L.L. Chang, L. Esaki and R. Tsu, Appl. Phys. Lett. 24 (1974) 593.
[2]   S. Hiyamizu, S. Muto, T. Inata, S. Sasa, T. Fujii, K. Imamura, H. Ohnishi and Yokoyama, Physics and Technology of Submicron Structures, Springer Series in Solid–State Science 83, eds. H. Heinrich, G. Bauer and F. Kuchar (1988) 287.
[3]   F. Capasso, Heterostructure Band Discontinuities: Physics and Device Applications, eds. F. Capasso and G. Margaritondo (North Holland, 1987) 339.
[4]   S. Luryi, Heterostructure Band Discontinuities: Physics and Device Applications, eds. F. Capasso and G. Margaritondo (North Holland, 1987) 489.
[5]   B.R. Snell, K.S. Chan, F.W. Sheard, L. Eaves, G.A. Toombs, D.K. Maude, J.C. Portal, S.J. Bass, P. Claxton, G. Hill and M.A. Pate, Phys. Rev. Lett. 59 (1987) 2806.

[6]   L. Eaves, E.S. Alves, T.J. Foster, M. Henini, O.H. Hughes, M.L. Leadbeater,
      F.W. Sheard, G.A. Toombs, K. Chan, A. Celeste, J.C. Portal, G. Hill and
      M.A. Pate, *Physics and Technology of Submicron Structures*, Springer Series
      in Solid–State Sciences 83, eds. H. Heinrich, G. Bauer, F. Kuchar (1988) 74.
[7]   E.S. Alves, M.L. Leadbeater, L. Eaves, M. Henini, O.H. Hughes, A. Celeste,
      J.C. Portal, G. Hill and M.A. Pate, *Proc. Fourth Int. Conf. on Superlattices,
      Microstructures and Microdevices* (Trieste, 1988); Superlattices and
      Microstructures 4 (1989) 527.
[8]   P. England, J.R. Hayes, M. Helm, J.P. Harbison, L.T. Florez and S.J. Allen,
      Jr., Appl. Phys. Lett. 54 (1989) 1469.
[9]   M. Henini, M.L. Leadbeater, E.S. Alves, L. Eaves and O.H. Hughes, J.
      Phys.: Condens. Matter 1 (1989) 3025.
[10] M.L. Leadbeater, , E.S. Alves, L. Eaves, M. Henini, O.H. Hughes, A.
      Celeste, J.C. Portal, G. Hill and M.A. Pate, J. Phys. Condens. Matter 1
      (1989) 4865.
[11] M.C. Payne, J. Phys. C: Solid State Phys. 19 (1986) 1145.
[12] T. Weil, and B. Vinter, Appl. Phys. Lett. 50 (1987) 1281.
[13] S. Luryi, Appl. Phys. Lett. 47 (1985) 490.
[14] M.L. Leadbeater, E.S. Alves, L. Eaves, M. Henini, O.H. Hughes, A. Celeste,
      J.C. Portal, G. Hill and M.A. Pate, Phys. Rev. B 39 (1989) 3438.
[15] B. Ricco and M.Ya. Azbel, Phys. Rev. B 29 (1984) 1970.
[16] V.J. Goldman, D.C. Tsui and J.E. Cunningham, Phys. Rev. Lett. 58 (1987)
      1257.
[17] T.C.L.G. Sollner, Phys. Rev. Lett. 59 (1987) 622.
[18] V.J. Goldman, D.C. Tsui and J.E. Cunningham, Phys. Rev. Lett. 59 (1987)
      623.
[19] G.A. Toombs, E.S. Alves, L. Eaves, T.J. Foster, M. Henini, O.H. Hughes,
      M.L. Leadbeater, C.A. Payling, F.W. Sheard, P.A. Claxton, G. Hill, M.A.
      Pate and J.C. Portal, *Proc. 14th Int. Symposium on Gallium Arsenide and
      Related Compounds* (Crete, 1987); Inst. of Physics Conf. Series 91 (1988)
      581.
[20] T.J. Foster, M.L. Leadbeater, L. Eaves, M. Henini, O.H. Hughes, C.A.
      Payling, F.W. Sheard, P.E. Simmonds, G.A. Toombs, G. Hill and M.A.
      Pate, Phys. Rev. B 39 (1989) 6205.
[21] E.S. Alves, L. Eaves, M. Henini, O.H. Hughes, M.L. Leadbeater, F.W.
      Sheard, G.A. Toombs, G. Hill and M.A. Pate, Electronics Lett. 24 (1988)
      1190.
[22] A. Zaslavsky, V.J. Goldman and D.C. Tsui, Appl. Phys. Lett. 53 (1989)
      1408.
[23] M.L. Leadbeater, E.S. Alves, L. Eaves, M. Henini, O.H. Hughes, F.W.
      Sheard and G.A. Toombs, Semicond. Sci. Technol. 3 (1988) 1060.
[24] F.W. Sheard, and G.A. Toombs, Appl. Phys. Lett. 52 (1988) 1228.
[25] V.J. Goldman, D.C. Tsui and J.E. Cunningham, Phys. Rev.B. 35 (1987)
      9387.
[26] C.A. Payling, E.S. Alves, L. Eaves, T.J. Foster, M. Henini, O.H. Hughes,
      P.E. Simmonds, J.C. Portal, G. Hill and M.A. Pate, *Proc. 3rd Int. Conf. on
      Modulated Semiconductor Structure* (Montpellier, France); J. Physique C5
      (1987) 289.
[27] C.A. Payling, E.S. Alves, T.J. Foster, M. Henini, O.H. Hughes, P.E.
      Simmonds, F.W. Sheard and G.A. Toombs, Surf. Science 196 (1988) 404.

[28] F.F. Fang, and W.E. Howard, Phys. Rev. Lett. 16 (1966) 797.

[29] L. Eaves, B.R. Snell, D.K. Maude, P.S.S. Guimaraes, D.C. Taylor, F.W. Sheard, G.A. Toombs, J.C. Portal, L. Dmowski, P. Claxton, G. Hill, M.A. Pate and S.J. Bass, *Proc. 18th Int. Conf. on Physics of Semiconductors* (Stockholm, 1986), ed. O. Engström (World Scientific, 1987) 1615.

[30] V.J. Goldman, D.C. Tsui and J.E. Cunningham, Solid State Electronics 31 (1988) 731.

[31] D.K. Roy, *Tunneling and Negative Resistance Phenomena in Semiconductors* (Pergamon Press, 1977).

[32] F.W. Sheard, and G.A. Toombs, *Proc. 5th Int. Conf. on Hot Carriers in Semiconductors*, to be published in Solid State Electronics (1989).

[33] E.E. Mendez, L. Esaki and W.I. Wang, Phys. Rev. B 33 (1986) 2893.

[34] E.E. Mendez, *Physics & Applications of Quantum Wells and Superlattices*, NATO ASI Series B; Physics 170 (1987) 159.

[35] L. Eaves, G.A. Toombs, F.W. Sheard, C.A. Payling, M.L. Leadbeater, E.S. Alves, T.J. Foster, P.E. Simmonds, M. Henini, O.H. Hughes, J.C. Portal, G. Hill and M.A. Pate, Appl. Phys. Lett. 52 (1988) 212.

[36] M. Tsuchiya, T. Matsusue and H. Sakaki, Phys. Rev. Lett. 59 (1987) 2356.

[37] D.G. Hayes, M.S. Skolnick, P.E. Simmonds, L. Eaves, D.P. Halliday, M.L. Leadbeater, M. Henini and O.H. Hughes, *Proc. 4th Int. Conf. on Modulated Semiconductor Structures*, to be published in Surface Science (1989).

[38] S. Sen, F. Cappasso, A.C. Gossard, R.A. Spah, A.L. Hutchinson and S.N.G. Chu, Appl. Phys. Lett. 51 (1987) 18.

[39] M. Heiblum, M.V. Fischetti, W.P. Dumke, D.J. Frank, I.M. Anderson, C.M. Knoedler and L. Osterling, Phys. Rev. Lett. 58 (1987) 816.

L. Eaves is with the Department of Physics, University of Nottingham, Nottingham NG7 2RD, England

ELECTRONS AS GUIDED WAVES IN LABORATORY STRUCTURES: STRENGTHS AND PROBLEMS

Rolf Landauer

Analogies between the Schrödinger equation and classical wave motion are deeply imbedded in the development of quantum mechanics. This is illustrated through the long history of the WKB approximation. We review the diversity and respective validity of expressions which relate the conductance of a sample to its transmissive behavior. The literature frequently alludes to supposed derivations of these results from linear response theory. We emphasize the difficulties faced by such derivations, and stress the relationship of electron reservoirs to radiative black bodies. The diversity of possible voltage probes is also emphasized.

1. INTRODUCTION

The analogy between quantum mechanical particle motion and classical wave motion is deeply imbedded in the origin of quantum mechanics through the work of De Broglie and Schrödinger, and is deeply imbedded in much of the development of quantum mechanics that followed immediately thereafter. The WKB approximation is only one of a number of possible examples and we will discuss it in some detail. Appreciation and use of the WKB approximation became widespread through quantum mechanics, but there were many earlier versions. A WKB approximation for Bessel functions was supplied by Carlini as early as 1817 [1]. In 1837 Green showed [2] that for tidal waves in a canal of slowly varying cross section the energy is transmitted without reflection. That already brings us very close to some of the subject matter of this book, to which we return later, in connection with Fig.1 of this concluding chapter. A WKB approximation for Legendre polynomials, $P_l(\cos\theta)$, was supplied by Laplace [3]. Gray and Schelkunoff [4] tell us that Liouville used the WKB approximation in 1837. Lord Rayleigh [5] in his *Theory of Sound* in 1896 remarks, "For example the amplitude of a sound wave moving upwards in the atmosphere may be determined by the condition that the energy remains unchanged." This is a physical description of the WKB approximation. By 1912 Lord Rayleigh [6] gave a detailed discussion of the WKB method. He includes a treatment of the total reflection occuring at a linear turning point, in terms of Bessel functions of order one–third. He also discusses the distributed reflections occurring for a wave propagating through a slowly varying medium. There are occasional modern references to a WKBJ or JWKB approximation which stress the 1923 publication by Harold Jeffreys [7]. That seems anti–climatic in view of the long earlier record, which we have cited incompletely.

W. van Haeringen and D. Lenstra (eds.), Analogies in Optics and Micro Electronics, 243–257.
© *1990 Kluwer Academic Publishers. Printed in the Netherlands.*

Further historical details are given in the book by Heading [8], who likes
"W.B.K.J." History is provided to a lesser extent by another monograph [9], which
prefers "JWKB".
      Later on we shall discuss, in detail, the relationship between the transmissive
behavior of a sample and its conductance. The WKB approximation, at energies
above all the potential peaks, allows complete transmission. Thus, the condition
for its applicability becomes a condition for complete transmission. The WKB
approximation can be considered to be the first term in a convergent series, in
which each term represents the reflections generated by the preceding term. Such
an expansion was first pointed out in connection with electromagnetic waves [10]
and then applied to the Schrödinger equation [11]. These concepts were revisited
by a number of subsequent authors and the technique is included in what Bellman
has called *Invariant Imbedding* [12]. Heinrichs [13] has, in turn, applied invariant
imbedding to modern electron localization problems.
      At a classical turning point the WKB approximation diverges, and *connection
formulae* are then invoked to relate the approximations on the two sides of the
turning point. Away from a turning point, in a classically allowed region, the
WKB approximation gives us wave motion with variable amplitude, but without
reflections. Such unreflected wave motion does not give us the band structure for
electrons in a periodic potential; Bragg reflections are needed for that. But the
WKB approximation supplemented with the connection formulae, for energies
below the potential peak of a periodic potential, does yield band structure [14].
That point, however, was already understood in 1924, before wave mechanics [15].
The analogy between classical waves in periodic arrays and electrons in crystals
was again emphasized in Brillouin's beautiful book *Wave Propagation in Periodic
Structures* [16]. The scene is revisited in this volume by the remarkable
experiments of E. Yablonovitch [17].
      The analogy between classical waves and electrons under the Schrödinger
equation is manifested in many other ways. One more example: Multiple scattering
in ordered and disordered media was discussed from a unified viewpoint for
electrons and classical waves by Lax [18]. It is, in fact, remarkable that the
theoretical work on electrons in disordered media has had only minor and
incidental contact with the work on electromagnetic waves in turbulent
atmospheres, or the work on sound wave propagation in ocean water. By and large,
these subjects have gone their separate ways.
      What is new then? Coherent motion of waves through an entire structure, as in
an organ pipe or waveguide, is analogous to the coherent motion of the electron
waves giving rise to quantization in atoms and molecules. Electrons traversing
manufactured structures, however, such as transistors, typically suffer inelastic
events. The emerging waves are incoherent with the incident waves. Only in recent
years have we made structures which are small enough and clean enough so that
the carriers have had a chance to show coherent motion in their transit through the
whole structure. This achievement in miniaturization, a byproduct of
microelectronic technology, has created the field celebrated in this volume. While
much of the new and exciting work deals with electronic waves *through* structures,
the study of quantized states in *closed* structures *without leads* has also been
undertaken. Started in Ref. [19], and extended by a number of investigators [20], it
is represented in this volume by the contribution of the editors [21].
      One subject not treated in this volume relates to tunneling. Its optical analog:
Evanescent waves in a low refractive index layer between optically denser regions

[22]. Light incident from one side, at an angle that would produce complete internal reflection if the low index material were semi–infinite, will be transmitted with exponential attenuation through a sandwich containing the low index material in the middle. Control of the refractive index in the middle layer can be used as a tool for modulating light [23]. In the case of quantum mechanical tunneling we have, in the last decade, learned how to evaluate the time that an electron spends traversing the tunneling barrier [24]. Similar questions can be raised for frustrated internal reflection, i.e. for the exponentially small optical transmission. They can be answered by the techniques applied to the quantum mechanical tunneling case [25], though some murky questions remain to be resolved in that connection.

The analogy between *high energy* electrons in vacuum, and optical waves, is stressed by the very word: electron optics. Admittedly, in much of that field we deal with classical electron motion; the analogy is to ray optics. But not always; one very impressive modern contribution comes in the form of electron holography [26].

A similar distinction between real wave propagation effects and more classical motion can be made in the recent mesoscopic work. Beenakker and van Houten [27] show us that many of the novel observed experiments can be explained via a classical billiard ball model. But certainly not all of them! The quantized resistanced observed in the quantum Hall effect and in the striking constriction conductance measurements discussed by the same authors in this volume [28] are inevitably and genuinely quantum mechanical! The analogy between electrons moving through potential fields in two–dimensional semiconductor structures and optical lenses has also been studied [29].

1.1   Electrons Are Not Photons

The analogies have severe limitations, and the brevity of this section should not be used to measure the importance of that point. Electrons, at least within low energy physics, are conserved, whereas photons can come and go. Electrons are generally assumed to obey a linear wave equation, aside from some advanced recent speculations [30]. Non–linear optics, on the other hand, is a flourishing field with a great many successful applications. Indeed, nonlinear electromagnetic wave propagation was studied long before the advent of the laser, starting with a 1923 paper by Salinger [31]. (This author, in an unpublished note, "A Personal View of Nonlinear Electromagnetic Wave Propagation," reviews this history and chides the quantum electronics community for its neglect of Salinger's work, and neglect of other papers that preceded the laser). These nonlinearities can be considered to be photon–photon interactions. Electrons, of course, can interact much more directly and effectively through their Coulomb charge. That is what causes electron devices such as the vacuum tube and the transistor to be so effective. The attempt to perform similar gating functions through nonlinear optics has had very limited success [32]. A Fabry–Pérot interferometer can be made to be dependent on the degree of excitation through nonlinear optical effects [33], i.e. through the dependence of the index of refraction on the electric field intensity. Resonant tunneling is the electron analog of the Fabry–Pérot. The existence, or not, of electrons trapped in such a structure can have a pronounced effect on the potential seen by other electrons, as shown in Ref. [34] and that effect is represented here by the contribution of L. Eaves [35].

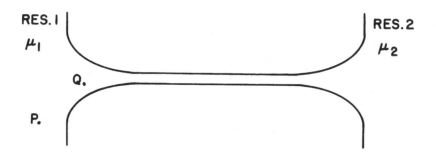

Fig. 1   Two reservoirs on each side of a perfect tube, at different
          electrochemical potentials μ_1 and μ_2. The tube is connected to the
          reservoirs in such a way that all carriers coming out of the tube pass
          into the reservoir without reflection.

In this somewhat pessimistic section, we will include one area where we believe
the analogy does hold. This author believes that devices depending on totally
· coherent transmission, whether of light waves or of electrons, are not a promising
replacement for transistor circuits as a tool for executing logic in a computer. That
case has been made elsewhere in some detail [32,36] and will not be repeated here.

2.   CONDUCTANCE BETWEEN RESERVOIRS

The fact that the conductance of a localized tunneling barrier can be calculated
from its transmissive behavior was understood by Frenkel in 1930 [37]. Frenkel's
result with a conductance proportional to transmission is valid only for small
transmission probability [38]. Nevertheless, written down decades before
convincing and controlled solid state tunneling was observed, it represented a
remarkable pioneering insight. Only within the past decade has there been
widespread appreciation that the conductance of a whole sample, not just of a
single barrier, could be calculated in a similar way. Even in this area the analog to
classical wave transport is apparent; in heat flow, we refer to the Kapitza
resistance related to the transmission of sound waves through an interface.
      Consider the sitiuation shown in Fig.1. We have reservoirs on the left and on
the right. Deep inside the reservoirs we are very close to thermal equilibrium with
the respective electrochemical potentials μ_1 and μ_2. The cylindrical part of the tube
is, for the moment, assumed to be uniform and causes no scattering. The
connection between the tube and the reservoirs is made through slowly flared
horns, so that waves coming out of the tube pass into the adjacent reservoir
without reflection.[1] Detailed balance applied to the thermal equilibrium case

---

[1]Any "matching" scheme which lets all the waves coming out of the tube move
into the reservoir will do, the adiabatic connection is not essential. Therefore we
ignore the question: Can a real geometry of limited length be adiabatic?

assures us, then, that all of the states coming out of the left hand reservoir and into the tube, up to the level μ_1, must be occupied. Let us assume that the tube allows N_T different quantized transverse states, below the Fermi level, each with a two–fold spin degeneracy. We are concerned here with linear conductivity; μ_1 and μ_2 are close to each other. In that case, elementary considerations, given in many papers, particularly in Ref. [39], lead to the conclusion that the resulting current is

$$j = N_T(2e^2/h)(V_1 - V_2), \qquad (1)$$

where $eV_1 = \mu_1$ and $eV_2 = \mu_2$. To eliminate distracting minus signs we have assumed transport by positive carriers of charge e. Alternatively, the reader may want to assign a minus sign to our charge e. All N_T channels contribute equally to the current flow in (1). Now, if the tube in Fig.1 causes some scattering of electrons, and if the probability of transmission, averaged over the N_T channels is T, then the current flow is reduced accordingly and the conductance becomes

$$G = TN_T(2e^2/h). \qquad (2)$$

We have, here, assumed that a single value of T applies to all relevant incident electrons near the Fermi–surface, regardless of their energy. The case where $k\theta$ (θ is the temperature) is high enough to invalidate this will be discussed later.

Eqs. (1) and (2) are widely known, and widely used. Unfortunately, the conditions for their validity are not equally well appreciated.

First the good part. The exact nature of the scattering processes which give rise to the transmission probability T is irrelevant. The scattering need not be elastic. The scattering can also include effects depending on the electron spin, or effects dependent on sophisticated many–body interactions. In the case where inelastic scattering is present, admittedly, Eq. (2) has a limited range of utility. It does assume that there is a single transmission coefficient applicable to all the relevant incident energies near the Fermi–level. Thus, $k\theta$ must be small compared to the energy range over which T varies appreciably. But, at sufficiently low temperatures, the incident electrons cannot lose much energy and the lattice vibrations do not have much energy to supply to the incident electrons. That limits inelastic scattering severely. We return to the case where the dependence of T on energy has to be taken into account, subsequently.

A number of authors have deduced conductance expressions with inelastic effects present [40]. This work is not irrelevant. In the case of (2) applied to elastic scattering, we do not get the answer immediately from (2); we must calculate T first from a microscopic model. Similarly for the inelastic case. Such calculations, however, should not be regarded as extensions or generalizations of (2); that equation is valid as it stands and does not need generalization to the inelastic case.

Now for the bad part in connection with (1) and (2). The reservoirs are essential, the expressions characterize a potential drop, or electrochemical potential drop, between points way inside the respective reservoirs. Within the conducting pipes we have a current flow, we do not have an electron population which is in thermal equilibrium. We can approach equilibrium only in the reservoir at a point sufficiently far into it, such that most of the carriers there have had a chance to come into equilibrium with the chemical potential of the reservoir. And far enough into the reservoir so that the density of carriers present there is characteristic of μ_1 with little influence from μ_2. At Q, for example, in Fig.1, we

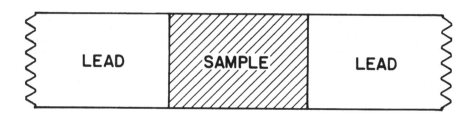

Fig. 2.   Scattering sample connected to leads which are of uniform width and
           semi–infinite.

will see a carrier population that has come to a substantial extent out of the tube,
and in part from reservoir 2. Further into the reservoir, at P, we will be
approaching a carrier population characteristic of μ_1. Of course, to equilibrate the
carriers arriving in the reservoir we need some inelastic scattering. It is, however,
the geometrical spreading which reduces the current density as we go more deeply
into the reservoir; and this is the really critical part. Payne [41] has emphasized
closely related views relating to the geometry of the reservoir.

Consider the particular case of Fig.1, with $T = 1$. Now assume that we make
measurements at two points along the narrow conductor, with two identical and
non–perturbative probes. (The existence of non–perturbative probes is defended in
Ref.[42] and more briefly in Sec.4 of this paper.) Assume that the two points are
sufficiently far from the reservoirs, so that any higher order evanescent modes (i.e.
modes whose transverse energy component is above that of the Fermi–level)
reaching in from the reservoirs have disappeared. Then at two arbitrary points
inside the conductor we will see identical conditions. Any non–perturbative
measurement of carrier population, or of electrostatic potential will be identical at
the two points; *there is no voltage drop along a perfect conductor*. We can, of
course, create one, by invoking a probe on the left which equilibrates only with the
carriers from the left, and a probe on the right which equilibrates only with the
carriers from the right. That is unreasonable.

The geometrical dilution provided by the reservoir symbolized in Fig.1 is
essential. A number of papers over the years, some of them cited in Ref.[42], claim
to have derived (2) from, or else apply it to, the geometry of Fig.2 invoking a
conductor of uniform width. In such a structure, of course, continuity of current
must be maintained. Nowhere, along the "leads" can we find the equilibrium that
is characteristic of a point deep inside a reservoir. Indeed, if we provide inelastic
scattering in the leads, far from the scattering region of interest, the potential will
have to change linearly with distance. Where, then, are the exact points between
which (2) applies?

In my own papers [43,44] I have always emphasized the analogy of an electron
reservoir to a black body, even though I did not understand the specifications of a
reservoir fully, before Imry's clarification [39]. What is a black body in radiation
theory? Typically, it is taken to be a *large* cavity with a *small* hole, from which

equilibrium radiation emanates. This sort of *geometrical spread is essential.* Furthermore, the "matching" is also essential. Clearly if the interface to the reservoir adds extra scattering, then the potential difference between reservoirs cannot possibly be predicted by a formula which ignores that effect.

We note that all the existing attempts to derive (2) from the Kubo formula utilize the geometry of Fig.2, and therefore cannot do what they claim to do. Indeed, if done self–consistently, with due attention to the Coulomb interactions between carriers [38], they yield the results we will quote in Sec.3, rather than (2). A recent version of these discussions [45] specifically tells us that it treats an "...arbitrary shape with an arbitrary static potential connected to N_L perfectly ordered, *straight infinite leads*" (italics added for emphasis). Ref.[45] does, later on, discuss reservoirs. But it is an auxiliary discussion, unrelated to the paper's formalism; the wave functions are not followed in or out of a reservoir. As we have emphasized, however, (2) characterizes a potential difference between reservoirs. I am most impressed by a quantum mechanical linear response theory which can calculate the potential difference between two points without concern about the behavior of the relevant states at, or near, these two points. That, of course, can be remedied. A linear response theory can and should allow for the proper geometry.

The approximation that is introduced by applying (2) to a reservoir of limited width, rather than infinite width is discussed in detail in Refs.[41] and [42].

2.1   Energy–Dependent Transmission

Now let us turn to the case where the transmission probability, averaged over incident channels, varies sufficiently rapidly with incident energy so that within the relevant energy range, $\sim k\theta$, at the Fermi–level, this variation cannot be ignored [46]. The rate at which current arrives from the left, in the energy range dE, is

$$j_{dE} = \sum_i (dn_{Ri}/dE)(ev_i)f(E). \tag{3}$$

This is a sum over incident channels, dn_{Ri}/dE is the density of carriers moving to the right in the ith channel, v_i is the velocity component in the direction of the current, for the ith channel, and f is the Fermi–occupation probability. In the usual fashion, $v_i(dn_{Ri}/dE) = 2/h$. Thus,

$$j_{dE} = \frac{2}{h} N_T ef dE. \tag{4}$$

Let us now, for the moment, concentrate on the case of a sample with purely elastic scattering. In the presence of a voltage bias V_{12}

$$(f_1 - f_2) = (-df/dE)eV_{12}. \tag{5}$$

This gives an excess transmitted current, within dE, of

$$j_{T,dE} = \frac{2}{h} N_T e^2 V_{12} T(E)(-df/dE) dE. \tag{6}$$

After integrating over E the resulting conductance becomes

$$G = \frac{2e^2}{h} N_T \int T(E)(-df/dE)dE, \tag{7}$$

where $(-df/dE) = f(1-f)/k\theta$ peaks at the Fermi–level. We might, incidentally, note that a given incident l.h.s. channel excites the various emerging, r.h.s. channels in a coherent fashion. But in calculating current flow, or occupation probability over an extended part of space, these phase relationships cease to matter.

Now let us consider the validity of (7) in the presence of inelastic scattering. Eq.(5) is then no longer applicable. Nevertheless, in the absence of an applied voltage there will be detailed balance. The current incident from the left in a given small energy range, and in a particular channel, and transmitted to a particular emerging channel on the right, will be balanced by the transport in the opposite direction. In the presence of an applied voltage there will be excess flux arriving from the left. This will be transmitted in the same proportions as the equilibrium portion of the incident flux. Thus, Eq.(7) still applies.

Some readers may be squeamish about the role of the Pauli principle. Does not the transmission process from left to right, to a particular channel, depend also on the extent to which the reflection process puts carriers incident from the right, into the same channel? The irrelevance of the Pauli principle, to this kind of question, can be demonstrated, for example, through the kind of wave packet analysis given in Ref.[47].

The case where the scatterer includes externally applied oscillating potential components is more complex. That can include, for example, traveling sinusoidal waves familiar from traveling wave tubes, linear accelerators, and the acousto–electric effect. But even in the time–dependent potential case it is likely that (2) still has a wide range of applicability..

2.2   History

The lingering confusion about the region of applicability of (2) has also served to obscure the proper assignment of credit for its origins. In this historical discussion we are only concerned with (2) and not with other expressions for conductance with different realms of applicability, taken up later. First of all, discussions based on the geometry of Fig.2 simply cannot be correct derivations of (2). The geometrical spread of the reservoir, and the "matching" to it, are essential. The need for the geometrical spread, possibly implicit in some earlier discussions, first became clear, explicit, and unambiguous in Ref.[39]. The generalization of (2) to the case of more than two reservoirs, universally used in the interpretation of experiments, including the quantized Hall effect, is due to M. Büttiker [48], and reviewed in Ref.[49]. In the most prevalent experimental configuration the leads used for introducing current are separate from those used for voltage measurements, and Büttiker's formulation becomes essential. The road toward that treatment started with Ref.[50] which introduced a reservoir attached via a lead carrying no net current, to a mesoscopic transport system. In Ref.[50] the added floating reservoir served as a source of inelastic processes. Later, it was recognized that the dangling reservoir, whose electrochemical potential adjusts so as to yield zero net current out of it, is a model of a voltage probe [48].

In a book devoted to analogies it is incumbent on us to cite the work of Erdös and collaborators, initiated in Ref.[51] and refined in Ref.[52], treating thermal transport. Erdös is concerned with transport through a system of limited extent, not with the prevalent thermodynamic limit. His boundary conditions are determined by phonon fluxes emanating from blackbody reservoirs. His temperature, along the conducting body, is measured by a blackbody reservoir coming into equilibrium with the transport system. It deviates from the approach that we have stressed in only one aspect: Erdös does not inquire about the correlation of the phonon flux and/or temperature gradient with the location of the scatterers. He is basically concerned with an ensemble average, not with the behavior of a sample with a particular distribution of phonon scatterers.

3.   THERE ARE MANY CONDUCTANCE FORMULAS

The original version of the conductance formula put forth by this author [43] was

$$G = (2e^2/h)T/R \,, \tag{8}$$

for the one–dimensional case. A many–dimensional generalization was provided by Azbel [53] and analyzed in detail in Ref.[54]. A further generalization to the case where transmission depends appreciably on energy, within the range of interest near the Fermi–level, was provided in Ref.[55]. Much of the subsequent literature asks, unfortunately, whether (8) is right or whether the one–dimensional version of (2)

$$G = (2e^2/h)T \tag{9}$$

is correct. Similarly for the many–dimensional case. Both are, of course, valid in their respective realms of applicability. Indeed, these two results are only two out of a spectrum of many more.

Fig.3 shows the realm of applicability of (8) and its multi–dimensional versions. In contrast to (2) which characterizes the potential difference between reservoirs, Eq.(8) and its multi–dimensional generalization give us the potential difference between the perfectly conducting leads. The potential difference must be averaged over a number of Fermi wavelengths to remove the effects of interference oscillations. Eq.(8), just like (2), does not require detailed further assumptions about the mechanism that causes reflection. Detailed discussions for the case of a sequence of incoherent scatterers were given in Ref.[46] and Ref.[56].

Scattering action will cause an interference pattern between the incident wave and scattered or reflected waves. This results in oscillatory charge densities and oscillatory potentials, even after introducting self–consistency. Büttiker [57] has studied the effects of these oscillations on voltage measurements, assuming the availability of very high resolution probes. These interference oscillations, which contain some components with wavelenghts long compared to the Fermi wavelength, are familiar from electromigration theory [58]. Over distances which are large compared to the relevant wavelengths and large compared to the screening length, we must have charge neutrality. As we deviate from thermal equilibrium, and introduce a non–oscillatory component in the deviation in carrier density near the original Fermi–level, the bottom of the band, determined by the local potential, will have to move so as to reset the total local electronic charge

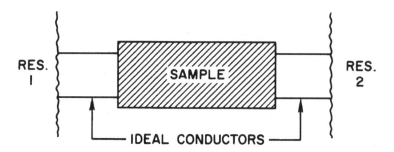

Fig. 3   Reservoirs connected to ideal conducting leads which are in turn
              connected to the sample with its scattering action.

density to that present originally. That is what is needed in the derivation of (8),
and its many–dimensional generalization. It is not invalidated by the existence of
interference oscillations.

The two results discussed so far are only two out a myriad of possible
situations, for samples with two contacts. In both Fig.1 and Fig.3 the electron
population being introduced into the sample is that characteristic of a reservoir at
thermal equilibrium. The electrons arriving at the scatterer, from the left, are all
in equilibrium with each other. In general, that need not be the case. The
electrons, for example, arriving at a sample through a typical ohmic conductor will
provide an incident carrier population characteristic of the shifted Fermi–sphere
typical of such conductors. The situation will be further complicated by the fact
that some of the carriers incident on the sample will have recently come from the
sample. Additional discussion of such problems is contained in Ref.[59], and more
briefly in Ref.[60].

We have already criticized supposed derivations of (2) from Kubo–type linear
response theory. Our earlier complaint was based on geometry. There is, however,
a deeper problem. As just emphasized, samples have contacts through which
carriers come and go, and the conductance depends on the distribution of the
arriving carriers in real space and in momentum space. Linear response theory,
however, by definition describes the response of a Hamiltonian system to a small
perturbation. That does not include systems which have places where carriers enter
and leave. Linear response theory cannot reproduce the diversity of results which
we can and need to get from the transmission viewpoint, and which reflect physical
reality. It is different for a closed loop without leads to the outside, and driven by
magnetic flux. That is a Hamiltonian system. But, as a Hamiltonian system, it
does not have a resistance [19]. Nevertheless, derivations of (8) from linear
response theory which start from a closed loop and pay attention to
self–consistency [61] need not be seriously flawed. They only share the common
difficulty of all attempts to describe dissipative effects from a Hamiltonian
viewpoint, without explicity describing the degrees of freedom which take up the
energy. Note, also, that in this discussion we have concentrated on the *linear*
response. Ref.[62] treats the behavior of electrons in closed loops, at higher fields.

Eq.(2) invokes the potential difference between terminating reservoirs. If we are concerned with the voltage distribution along a circuit, involving points outside a terminating reservoir, then (2) becomes irrelevant. Eq.(8) does give an expression for voltage drop along the circuit, but it too is valid only under special circumstances that we have discussed. Some experiments include a series resistance correction to allow for a voltage drop between a probe and a point at which the voltage is of interest. The celebrated quantized constriction experiments [63] are in this category. We note, as was already emphasized in Ref.[42], that such a series resistance correction immediately takes us away from anything that can be directly related to (2) or its multi–terminal generalizations [49].

When do we need to know the internal voltage distribution? We do not need to know it if we are interested in the linear response, between reservoirs [57]. We do need to know the spatial pattern of the voltage drop if we want to evaluate the nonlinear behavior [46,64]. In that case the voltage distribution becomes an essential part of the total scattering potential. We also need to know about spatial variations if we are interested in the force causing defect motion [65] in the presence of current flow. Spatial variations will be important for ac transport which can permit capacitive short–circuiting for the parts of space where transport is particularly hard. Spatial variations will matter if we measure spatial potential distribution in some way other than through reservoirs. And as already stated in this section, that includes the case where we correct experimental data for series resistance. Ref.[66] is a recent example of one of the few transport theory papers (in contrast to device related papers) which give serious attention to self–consistency.

4.   WHAT AND WHERE IS THE VOLTAGE?

Ref.[42] has emphasized the variety of possible voltage probes, and we briefly allude to that here. In the typical mesoscopic experiment we have voltage probe leads which are comparable to the sample. To many theoreticians a voltage, until a few years ago, was the integral of a field. The field, in turn, was something measured via a hypothetically small charge. Now, that fundamentally oriented theoreticians have discovered how it is really done in the typical experiment, they have enshrined the use of perturbative voltage leads as the only possible approach. In some of my lectures I refer to a "Probe Cult" which worships capricious entities that complicate human life. It is very contrary to the normal style in physics, where we look for the least perturbative method of measurement. We do not measure the temperature of a patient by bringing the patient into a cold house and then checking to see how much the house warms up.

We can, of course, measure electrochemical potentials, e.g. by letting a small reservoir equilibrate with some portion of a transport system, as in Ref.[49]. But we can also measure electrostatic voltages. Volta already recognized that dissimilar metals, in equilibrium with each other, have a potential difference. Lord Kelvin returned to this observation [67] and firmly established the measurement of contact potential differences. We can use such purely capacitive methods for the measurement of potential differences, we do not need to tie a conductor to the sample.

If we do use probes which actually couple conductively to the sample, the coupling can be very loose and noninvasive. The probe need not disturb the underlying conduction process appreciably [68]. Measurements with such loosely coupled probes may still be complicated by multiple scattering effects [49,69] but that does not make them *invasive*. Tunneling probes have been used to measure transport voltages [70], that is not just a speculative suggestion. The tunneling probes provide, presumably, an electrochemical potential measurement. But the tunneling tip is loosely coupled; it is not a seriously perturbative measurement.

The phase space of possible measurement techniques is very large; perhaps we can use electron beam holography [26]. The energy gain of electrons which have passed through a narrow constriction determines their velocity, and therefore their subsequent trajectory in a magnetic field. This was used in Ref.[71] to determine the voltage drop and therefore the resistance. No theory paper discussing voltage measurements had the imagination to anticipate this beautiful method. Undoubtedly there are more inventions like that one left to be made. Büttiker's minute phase sensitive voltage probes [57] give another example that illustrates the great variety of possible probes.

The typical linear response theory, which does not pay attention to self–consistency, discourages a concern with the spatial variation of the transport field. Linear response theory for mesoscopic samples has, to a large extent, been replaced by a multi–reservoir viewpoint [49] which concentrates on events at the reservoirs. That is its strength and simplicity but, unfortunately, it once again discourages a concern with spatial variations along the current path. The analogous work dealing with classical wave propagation in optics or microwaves was never beset by such a psychological bottleneck. Of course, if we want to do device kinetics, where we need to be concerned with nonlinear behavior, we must face the spatial variation questions. Similarly, we might expect every elementary solid state text to ask about the spatial variation of current and field near a localized point defect scatterer; none of them do. In fact the only text which to some extent reflects the viewpoint presented here, calculating conductances from transmissive behavior, is Ref.[72]. It is no coincidence that this is a device oriented book.

REFERENCES

[1]   G.N. Watson, *A Treatise on the Theory of Bessel Functions* (Cambridge University Press, 1946) p. 6.
[2]   H. & B.S. Jeffrey, *Methods of Mathematical Physics* (Cambridge University Press, 1946) p. 492.
[3]   Laplace, *Mécanique Céleste,* livre 11, supplément 1. [Reference given by G.N. Watson, Camb. Phil. Trans. 22 (1918) 290].
[4]   M.C. Gray and S.A. Schelkunoff, Bell Syst. Tech. J. 27 (1948) 350.
[5]   Lord Rayleigh, *The Theory of Sound,* Vol. II, (Macmillan, London, 1896) p. 68.
[6]   Lord Rayleigh, *On the Propagation of Waves through a Stratified Medium, with Special Reference to the Question of Reflections,* Proc. R. Soc. London, Ser. A 86 (1912) 207.
[7]   H. Jeffreys, Proc. London Math. Soc. 23 (1923) 428.
[8]   J. Heading, *An Introduction to Phase–Integral Methods* (Spottiswoode, Ballantyne, London, 1962).
[9]   M. Fröman and P.O. Fröman, *JWKB Approximation* (North Holland, Amsterdam, 1965).
[10]   H. Bremmer, Physica 15 (1949) 593.
[11]   R. Landauer, Phys. Rev. 82 (1951) 80.

[12] R. Bellman and G.M. Wing, *An Introduction to Invariant Imbedding* (John Wiley, New York, 1975).
[13] J. Heinrichs, Phys. Rev. B33 (1986) 5261.
[14] N.L. Balazs, Ann. Phys. 53 (1969) 421.
[15] H. Jeffreys, Proc. London Math. Soc. 23 (1924) 437.
[16] L. Brillouin, *Wave Propagation in Periodic Structures* (McGraw–Hill, New York, 1946).
[17] E. Yablonovitch, this volume, Chap. 8.
[18] M. Lax, Rev. Mod. Phys. 23 (1951) 287.
[19] M. Büttiker, Y. Imry, and R. Landauer, Phys. Lett. 96A (1983) 365.
[20] Y. Gefen and D.J. Thouless, Phys. Rev. Lett. 59 (1987) 1752; H–F. Cheung, Y. Gefen, E.K. Riedel, IBM J. Res. Dev. 32 (1988) 359; H–F. Cheung, Y. Gefen, E.K. Riedel and W–H. Shih, Phys. Rev. B37 (1988) 6050; O. Entin–Wohlman and Y. Gefen, Europhys. Letts. 8 (1989) 477; G. Blatter and D.A. Browne, Phys. Scripta T25 (1989) 353; E.K. Riedel, H–F. Cheung, Phys. Scripta T25 (1989) 357; H–F. Cheung and E.K. Riedel, Phys. Rev. B40 (1989) 9498; H. Bouchiat and G. Montambaux, J. Phys. France 50 (1989) 2695.
[21] D. Lenstra and W. van Haeringen, this volume, Chap. 1.
[22] S. Zhu, A.W. Yu, D. Hawley, and R. Roy, Am. J. Phys. 54 (1986) 601.
[23] A.H. Nethercot, *Electro–Optic Light Coupling of Optical Fibers*, U.S. Patent 3,208,342 (Sept. 28, 1965); A.W. Yu, S. Zhu and R. Roy, Appl. Opt. 24 (1985) 3610; R. Cuykendall, Appl. Opt. 27 (1988) 1772.
[24] R. Landauer, Nature 341 (1989) 567; R. Landauer and M. Büttiker, *Response to "The Büttiker–Landauer Model Generalized"* to be published J. Stat. Phys.
[25] M. Büttiker and R. Landauer, unpublished.
[26] S.M. Dambrot, Science 246 (1989) 31.
[27] C.W.J. Beenakker and H. van Houten, Phys. Rev. Lett. 63 (1989) 1857.
[28] H. van Houten and C.W.J. Beenakker, this volume, Chap. 13.
[29] U. Sivan, M. Heiblum and C.P. Umbach, *An Electrostatic Electron Lens in the Ballistic Regime*, preprint.
[30] R. Thompson, Nature 341 (1989) 571.
[31] H. Salinger, Arch. Elektrotech. 12 (1923) 268.
[32] R. Landauer, in the Proceedings of *Frontiers in Condensed Matter Physics*, Bar–Ilan University, Jan. 8–11, 1990 (Physica A, to be published).
[33] L.A. Lugiato, in *Progress in Optics*, Vol. 21, E. Wolf, ed. (North–Holland, Amsterdam, 1984) p.71.
[34] A. Zaslavsky, V.J. Goldman, D.C. Tsui and J.E. Cunningham, Appl. Phys. Lett. 53 (1988) 1408.
[35] L. Eaves, this volume, Chap. 14.
[36] R. Landauer in *Nanostructure Physics and Fabrication*, eds. M.A. Reed and W.P. Kirk (Academic Press, San Diego, 1989) p.17; R. Landauer, Phys. Today 42 (1989) 119.
[37] J. Frenkel, Phys. Rev. 36 (1930) 1604. I am indebted to M. Büttiker for drawing my attention to this paper. In the past, I have cited a later 1933 review paper.
[38] R. Landauer, Phys. Lett. 85A (1981) 91.
[39] Y. Imry, in *Directions in Condensed Matter Physics*, eds. G. Grinstein and G. Mazenko (World Scientific, Singapore, 1986) p.101.

[40]   Y. Gefen and G. Schön, Phys. Rev. B30 (1984) 7323; M. Büttiker, Phys.
       Rev. B33 (1986) 3020; S. Datta and M.J. McLennan in *Nanostructure
       Physics and Fabrication*, eds. M.A. Reed and W.P. Kirk (Academic Press,
       San Diego, 1989) p.241; S. Datta, Phys. Rev. B40 (1989) 5830; S. Feng,
       *Quantum Transport in the Presence of Dissipation: Generalized Landauer
       Formula*, preprint.
[41]   M.C. Payne, J. Phys. Condens. Matt. 1 (1989) 4931.
[42]   R. Landauer, J. Phys. Condens. Matt. 1 (1989) 8099.
[43]   R. Landauer, Philos. Mag. 21 (1970) 863.
[44]   R. Landauer, Z. Phys. B21 (1975) 247.
[45]   H.U. Baranger and A.D. Stone, Phys. Rev. B40 (1989) 8169.
[46]   P.F. Bagwell and T.P. Orlando, Phys. Rev. B40 (1989) 1456.
[47]   R. Landauer, Physica D38 (1989) 226.
[48]   M. Büttiker, Phys. Rev. Lett. 57 (1986) 1761.
[49]   M. Büttiker, IBM J. Res. Develop. 32 (1988) 317.
[50]   M. Büttiker, Phys. Rev. B32 (1985) 1846.
[51]   P. Erdös, Phys. Rev. A138 (1965) 1200.
[52]   P. Erdös, Phys. Rev. A139 (1965) 1249; P. Erdös and S.B. Haley, Phys. Rev.
       184 (1969) 951; P. Erdös, S.B. Haley, J.T. Marti and J. Mennig, J. Comput.
       Phys. 6 (1970) 29; P. Erdös and S.B. Haley, Phys. Rev. B4 (1971) 669.
[53]   M. Ya. Azbel, J. Phys. C14 (1981) L225.
[54]   M. Büttiker, Y. Imry, R. Landauer and S. Pinhas, Phys. Rev. B31 (1985)
       6207.
[55]   U. Sivan and Y. Imry, Phys. Rev. B33 (1986) 551.
[56]   R. Landauer and M. Büttiker, Phys. Rev. B36 (1987) 6255.
[57]   M. Büttiker, this volume, Chap. 12; Phys. Rev. B40 (1989) 3409.
[58]   C. Bosvieux and J. Friedel, J. Phys. Chem. Solids 23 (1962) 123; R.
       Landauer, J. Phys. C8 (1975) L389.
[59]   R. Landauer, in *Localization, Interaction and Transport Phenomena*, eds. B.
       Kramer, G. Bergmann and Y. Bruynseraede (Springer, Heidelberg, 1985)
       Vol. 61, p.38.
[60]   R. Landauer, Z. Phys. B–Condensed Matter 68 (1987) 217.
[61]   D.J. Thouless, Phys. Rev. Lett. 47 (1981) 972.
[62]   D. Lenstra and W. van Haeringen, Physica B128 (1985) 26.
[63]   B.J. van Wees, L.P. Kouwenhoven, E.M.M. Willems, C.J.P.M. Harmans,
       J.E. Mooij, H. van Houten, C.W.J. Beenakker, J.G. Williamson and C.T.
       Foxon, *Quantum Ballistic and Adiabatic Electron Transport, Studied with
       Quantum Point Contacts*, preprint.
[64]   D. Lenstra and R.T.M. Smokers, Physica B151 (1988) 503; R. Landauer in
       Non–linearity in Condensed Matter, eds. A.R. Bishop, D.K. Campbell, P.
       Kumar and S.E. Trullinger (Springer, Heidelberg, 1987) p.2.
[65]   R.S. Sorbello and C.S. Chu, IBM J. Res. Dev. 32 (1988) 58; A.H.
       Verbruggen, IBM J. Res. Dev. 32 (1988) 93.
[66]   I.B. Levinson, Sov. Phys. JETP 68 (1989) 1257.
[67]   Sir William Thomson, Lord Kelvin, *Mathematical and Physical Papers*, Vol.
       VI (Cambridge Univ. Press, Cambridge, 1911) pp. 110–147. I am indebted to
       P. Stiles for bringing Kelvin's item to my attention.
[68]   C.S. Chu and R.S. Sorbello, Phys. Rev. B40 (1989) 5950.
[69]   H. Baranger, Presentation at *Workshop on Quantum Electrical Engineering*,
       Wayzata, Minnesota (1988).

[70] J.R. Kirtley, S. Washburn and M.J. Brady, IBM J. Res. Dev. 32 (1988) 414;
      J. Kirtley, S. Washburn and M.J. Brady, Phys. Rev. Lett. 60 (1988) 1546;
      J.P. Pelz and R.H. Koch, Rev. Sci. Instrum. 60 (1989) 301.
[71] J.G. Williamson, H. van Houten, C.W.J. Beenakker, M.E.I. Broekaart,
      L.I.A. Spendeler, B.J. van Wees and C.T. Foxon, *Hot Electron Spectrometry
      with Quantum Point Contacts*, Phys. Rev. B. to be published.
[72] S. Datta, *Quantum Phenomena* (Addison–Wesley, Reading, 1989).

Rolf Landauer is with the IBM Research Division, T.J. Watson Research Center,
Yorktown Heights, New York 10598, USA.

INDEX